IEE MANAGEMENT OF TECHNOLOGY SERIES 22

Series Editor: J. Lorriman

Intellectual Property Rights for Engineers

2nd Edition

Other volumes in this series:

Volume 4 **Marketing for engineers** J. S. Bayliss
Volume 8 **Business for engineers** B. Twiss
Volume 15 **Forecasting for technologists and engineers** B. C. Twiss
Volume 17 **How to communicate in business** D. J. Silk
Volume 18 **Designing businesses: how to develop and lead a high technology company** G. Young
Volume 19 **Continuing professional development: a practical approach** J. Lorriman
Volume 20 **Skills development for engineers** K. Hoag
Volume 22 **Intellectual property for engineers 2nd Ed.** V. Irish

Intellectual Property Rights for Engineers, 2nd Edition

Vivien Irish

The Institution of Electrical Engineers

Published by: The Institution of Electrical Engineers, London,
United Kingdom

The Institution of Electrical Engineers,
Michael Faraday House,
Six Hills Way, Stevenage,
Herts., SG1 2AY, United Kingdom

www.iee.org

British Library Cataloguing in Publication Data
Irish, Vivien, 1942–
Intellectual property rights for engineers. – 2nd ed.
1. Intellectual property – Great Britian 2. Engineers – Great Britian
I. Title II. Institution of Electrical Engineers
346.4′1048

ISBN 0 86341 490 7

Typeset in India by Newgen Imaging Systems (P) Ltd., Chennai, India
Printed in the UK by MPG Books Ltd., Bodmin, Cornwall

Contents

Chapter 1
Introduction

Engineers are natural innovators, whether they are involved in the basic development of a completely new system, or in making an existing system work better.

Engineers and scientists[1] are frequently willing to share their skills and ideas with others. Unfortunately, they sometimes find that others are less than willing to give appropriate credit or share the profits that can be made from these skills and ideas. Having one's idea stolen is an unhappy situation, but not an infrequent one.

This is where the law comes in, like it or not – and many engineers do not like the concept of legal controls on ideas at all. But we all need salaries or consultancy fees. If our work does not result in saleable products or services, then we as engineers will suffer financially. Our employers and work-providers need to make profits, and legally protecting the innovative efforts of engineers is one way of assisting this. The innovator, be it individual or company, should benefit from the investment of time, effort and money.

What is good for the individual engineer and company is also good for the country as a whole. The value of innovation to the UK is often given an airing by politicians, and the latest financial support is set out in the Sainsbury Report of 2004, which provides information about a £50 million grant programme for technological innovation. The law to protect such innovation is already in place and a basic knowledge can provide a good foundation for making sure that innovative effort is put to appropriate use.

Many engineers regard legal protection for innovation with misgivings. They may have tried to find out what is required and been turned off by the difficulty of finding explanations at the appropriate level. Legal text books are not easy reading, even for lawyers.

The aim of this book is to show that the relevant law is not as difficult as is sometimes imagined, certainly in its general application. The minutiae can be left

[1] This book was written with engineers in mind, but applies to scientists also, if they will excuse the avoidance of clumsy terminology and interpret the references to 'engineers' as meaning 'engineers and scientists'.

to the legal professionals who love to analyse the fine detail. The knowledge of the overall principles an engineer needs in order to use intellectual property (IP) law need not be deep, and can even be limited to a recognition of when to seek legal advice. This book goes further than that step, although it is far from being a legal text book. It is written on the basis that engineers generally like to know the reason for doing something. It tries to explain the general principles of the law protecting innovation, without going into great detail or giving all the exceptions to a general rule. It quotes the law and legal cases only if the author thinks this will help to make a point clear. The book therefore gives only an outline and general guidance, and expert advice is still essential in many cases.

The general term for the legal rights protecting innovation is intellectual property rights (IPRs). The phrase has become better known in the last 10 years or so, as the value of IPRs is becoming recognised by management and writers on management, in addition to government initiatives.

In the UK there are six basic IPRs – patents, copyright, registered designs, design right, trade marks and confidential information. Some of these terms, such as patents, copyright and trade marks, will be familiar, others less so: design right was introduced in August 1989, as was topography right, a special form of design right. Some types of rights have applications both to engineering and well beyond it, such as copyright and confidential information. Trade marks are somewhat peripheral to most engineering work, but are included because use of trade marks helps manufacturers to sell their products and service providers to attract customers. Throughout the book there will be brief references to non-engineering topics, such as music and film-making, but the explanations are largely based on technology.

The rights often overlap. A consumer product might be protected by all six rights at some stage in its design, manufacture and marketing. Different rights last for different lengths of time. Some of them are based on statute law, when reference can be made to the law for definitions, others on common law, based on judges' decisions over years or decades, when a reference to case law is the only way to understand the principles.

This book is structured to explain the six rights in Chapters 2–6 (registered design, design right and topography rights are grouped together in Chapter 3), setting out what each right protects, how to ensure it applies and the extent to which it can be used to control the activities of other companies. Chapter 7 considers in depth who owns the rights and Chapter 8 looks at how their use is constrained by EU law. Chapter 9 covers licensing, that is, allowing others to use the right in return for payment, and how to sue if necessary; litigation is rare in the IP field but always generates interest so it is included for completeness. The final chapter summarises various additional aspects, such as company policies on protection of innovation, and sets out a few situations when an engineering manager might need to make decisions about IPRs.

To find the chapter appropriate to a particular type of engineering product or material, please refer to Table 1.1. Each type of product or material may be covered by several different types of IPRs and several chapters may therefore need to be consulted.

Table 1.1 Protection available

Chapter number	2	3			4	5	6
Chapter title	Copyright	Design right	Registered design	Topography right	Patents	Confidential information	Trade marks
Technical report or specification	✓	—	—	—	✓	✓	—
Engineering drawing on paper	✓	✓	✓	✓	✓	✓	—
Engineering drawing on screen	✓	✓	✓	✓	✓	✓	—
Electrical device	—	✓	✓	✓	✓	—	—
Mechanical device	—	✓	✓	—	✓	—	—
Computer program	✓	—	—	✓	✓	✓	—
Computer icon	—	—	✓	—	—	—	—
Manufacturing method	—	—	—	—	✓	✓	—
Test method	—	—	—	—	✓	✓	—
Name of a product	—	—	—	—	—	—	✓
Pictorial design or logo	✓	—	✓	—	—	—	✓

Because the laws that give different types of protection are different in structure, it has not been possible to follow a set format for each chapter although the basic aspects are always present. There is inevitably some repetition for clarity, such as brief comments on ownership and suing for infringement in each of Chapters 2–6, while greater detail on both these topics can be found later in the book.

Some aspects are inevitably interwoven. EU law affects use of IPRs and the Commission in Brussels continually issues directives relating to IPRs. The basic principles enshrined in the Treaty of Rome, especially the aspects relating to control of anti-competitive agreements and abuse of dominant positions, are explained in a separate chapter. Similar principles apply in the USA and Japan. Directives affecting particular IPRs are considered within the relevant chapter.

The examples, whether to illustrate the separate IPRs or to illustrate legal points from case law, have been chosen to have an engineering interest as far as possible. Unfortunately, the next dispute to be heard may alter the legal interpretation and there are references here and there to 'current interpretation' or similar phrases. Also, in the UK, the wording of the law is all important, not the intention of parliament when the law was passed. Interpretation can only be fairly certain if a legal term has been considered by the courts, and only really certain if the House of Lords has decided a particular point. Readers may find this irritating, but the law is often irritating to engineers and scientists used to greater certainty and predictability.

Some chapters contain references to the criminal law because some types of misuse of IPRs are criminal offences. Such misuses are closer to daily engineering activity than the reader might at first think – the use of software is very common and its misuse by copying is extremely easy. The position is explained in Chapter 2.

There is no such thing as British law. One body of law relates to England and Wales, and usually Northern Ireland: Scottish law is quite separate. In the IPR area, there are few practical differences although Scotland has separate courts and different legal terminology. In general, references to English law and the English courts will usually extend to the whole of the UK, in principle at least.

The assumption throughout is that the engineer generating an innovation is a UK national working in the UK. Some IPRs, especially patents and copyright, are similar in other countries and references are sometimes made to the position in important countries, such as the USA and Japan. Other legal areas, particularly designs and the law relating to confidentiality, may be very different, and little attempt is made to compare the position overseas. The EU is attempting to harmonise the law on IPRs throughout all common market countries but has some way to go.

TRIPS (Trade-Related aspects of Intellectual Property Rights) is a set of letters widely used in this book. Arising from the Uruguay Round of GATT, the General Agreement on Tariffs and Trade, the TRIPS Agreement came into effect in January 1995. It includes copyright, industrial designs, patents, layout of integrated circuits and undisclosed information including trade secrets. It applies to Member Countries of the World Trade Organisation.

The agreement covers the minimum standards of protection to be provided for each type of IPR, and also general principles applicable to IPR enforcement procedures. It prohibits discrimination against non-nationals and forbids some countries being given more favourable treatment than others. All Member Countries must bring their law into compliance. While many developed countries already comply, the developing world has a longer period to phase in new arrangements.

Each chapter will give a summary of the relevant TRIPS minimum provisions.

Closer to home, the European Patent Office (EPO) has allowed a single patent application to extend to all 25 of its members. NB: The EPO is NOT a European Community organisation but the membership is a close match, see Table 4.2. The Community Trade Mark Office in Spain is an EU organisation which allows a single trade mark application to apply to all EU states, and the same applies to Community Registered Design and Community Unregistered Design Right. Negotiations on a Community Patent failed in 2004, perhaps fatally.

The law is continually updated, either by new statute law, by amendments to comply with EU directives, or by the decisions of the courts. There is never a perfect time to write a general legal book, it will always be out of date almost before it is published. This book was written when no major changes were in view, so the timing is reasonably apt.

The legal protection of innovation by IPRs is not limited to giant steps forward on the same scale as those made by Michael Faraday or Isambard Kingdom Brunel. Most engineers make improvements to existing equipment. Something that can seem trivial from the engineering point of view can be very valuable commercially, and therefore worthy of protection. Engineers are often too modest about their own ideas; the author believes that all innovation is worth at least considering for protection. IP law and the rules that apply to it should never be a hindrance to an engineer's work. Nor need they be a hindrance if the engineer has a little knowledge of them. This book sets out to provide that knowledge.

Chapter 2
Copyright

2.1 Introduction and law

Every engineer reading this book is a copyright owner and every engineering company also owns copyright. The reason is that this particular legal right applies automatically to an immense range of material.

Copyright is associated with every literary, dramatic, musical or artistic work, to sound recordings, films, broadcasts and cable programmes. The implications for the engineer are clear for the items in the second part of the list, but the definition of 'literary work' is sufficiently broad to cover engineering reports and specifications, and computer software; engineering drawings are classified as 'artistic works'.

The legal right comes into effect automatically, there is no need to register it or take any action – there is nowhere in the UK that copyright can be registered.

Copyright does not protect an idea or a concept, it protects the way in which the idea is expressed, the precise words or the actual drawing. There is no test for literary or artistic merit but the work must be original; it must be created by the engineer and not copied from something else, and the creator must have contributed skill or labour.

In this age of the Internet and rapidly expanding use and misuse of digital copyright material, international cooperation is essential. This area of IP law is highly active and subject to change in accordance with international and EU agreements.

The main law applicable in the UK is the Copyright, Designs and Patents Act 1988, which will be referred to as the 88 Act. It was amended in the 1990s to meet EU requirements and will be amended again. A WIPO (World Intellectual Property Organisation) Treaty on Copyright in 1996 led to two EU directives on e-commerce and digital copyright. Until both are implemented in all EU countries, the UK is unlikely to change its law, but it will do so. Meanwhile, a High Court judge has referred to a directive as the basis for his decision. On that basis, this chapter is written as though both directives are fully in force.

2.2 Types of copyright

2.2.1 Literary and artistic copyright

2.2.1.1 Literary copyright

As soon as an engineer writes a report or a specification, a handbook or even a business letter, it is automatically protected by copyright. In the 88 Act, the relevant legal definition for a literary work is 'any work . . . which is written, spoken or sung', so any string of words qualifies as a literary work.

Copyright applies from the instant the work is recorded and the law refers to this being 'in writing or otherwise'. There is a broad definition of writing to cover every conceivable recording method and use of the word 'otherwise' means that the spoken word recorded on tape etc. is also included.

The 88 Act is silent on quality or literary merit. Furthermore, it does not require any action to be taken; copyright applies automatically from the moment the engineer puts words onto paper or screen. In either case the work has been recorded. The record does not have to be a permanent one, it just has to come into existence.

The author's intention is also irrelevant. An internal memo is legally protected to the same degree as a fully revised paper written for publication, although the latter will be a more valuable copyright.

The only hint of a criterion is on length; one word is not sufficient to be a copyright work. This was decided when Exxon Corporation, the oil company, took legal action against an insurance consultancy which began to use the same name. Exxon argued that it owned copyright in its name but the court held that a single word was not copyright, even when it was not a normal word, but made up specially as a company name. Titles of books and magazines, particularly when they consist of conventional words, are not copyright either.

The words do not have to be in a normal language; a list of code words or shorthand symbols will also be protected by copyright.

2.2.1.2 Copyright in computer programs

Computer programs are classified as literary works, the 88 Act is quite specific on this. A moment's thought shows that this is a reasonable position, since programs have always been referred to as written in a computer language and the code is arranged in a manner equivalent to sentences, paragraphs and chapters.

The 88 Act does not define 'computer program' or even 'computer'. When the new law was being debated, parliament decided not to include such definitions. The aim was to ensure that the law will still apply even when technology moves on and new types of machine are developed which are not recognisable as computers, or new methods of instructing hardware are devised.

A literary work must be recorded in writing before it is protected and the 88 Act defines this as 'any form of notation or code, whether by hand or otherwise and regardless of the method by which, or medium in or on which, it is recorded'. The clear intention is to cover all possible types of program record.

In contrast with conventional literary works where mere 'sweat of the brow' is good enough to attract legal protection, this is not enough for electronic copyright; there must be a spark of creativity to comply with EU regulations. The size of this spark is not yet known but it is unlikely to be large and it is difficult to see how any new program could be excluded. For bug-fixing changes, the former copyright will apply.

2.2.1.3 Copyright in databases and compilations

Copyright protection also extends to compilations, for example, a list of names and addresses in a telephone directory or a parts list. The main requirement is originality, that is, that the compilation has not been copied from elsewhere. Individual items may appear in other records but the selection and arrangement in the particular compilation will be sufficient to attract copyright protection – 'sweat of the brow' is good enough here.

At the time the 88 Act was drafted, 'compilation' was considered to apply to electronic databases but this changed in March 1996 when a new, *sui generis*, database right came into force, different from copyright but considered here for convenience. Since that date, for a compilation held on a computer as a database, there must be minimal creativity, the author's own intellectual input is needed, applied to the selection or arrangement of contents. Mere alphabetical listing is excluded. There must be substantial investment in obtaining or verifying the contents.

If the test is met, there will probably be two rights. The first will lie in the program that determines the structure of the database and the way in which files within it are organised and accessed and will be conventional software copyright. The second concerns how the material is entered into the database. If the entry can be made by the user, for example, in a small business database for invoicing or mail merge, then database right will be owned by that person, because that person has selected the material and the way in which it has been entered.

Alternatively, the database itself may be created with a view to its use on a commercial basis. An example is the IEE's database INSPEC, which gives access to published papers in the fields of electronics, computing and physics; it contains abstracts of over 8 million scientific and technical papers and increases by over 400 000 records a year. IEE owns the legal rights.

2.2.1.4 The spoken word

Readers who make speeches should know that their spoken words count as literary copyright, even if they are not speaking from a fully prepared text but from notes or even extemporaneously. The recording of the speech necessary to attract copyright can be in writing, such as in shorthand, or otherwise, such as on tape or disk. An important factor is that the speaker owns the copyright in the record even if he or she does not make or authorise the recording. There may be other copyrights involved, so that a shorthand writer will have copyright in a particular shorthand version and the sound recordist will own copyright in the tape, but the speaker will have overall control through copyright in the words which have been recorded.

2.2.1.5 Artistic copyright

Artistic copyright is relevant to engineers because it applies to technical drawings, photographs and to designs of buildings and other structures. Copyright protection extends to this type of material irrespective of artistic quality, although it must be original, that is, not copied from elsewhere.

The general term 'graphic works' applies to any drawing, diagram, map, chart or plan. This means that any workshop drawing or rough sketch is protected. In the 1970s there were several legal decisions in which copyright was held to apply to drawings of a car engine, a gear box and an exhaust system.

A work of architecture is protected if it is a building, which includes any fixed structure. This probably applies to an overhead line tower or a large, fixed transformer as well as more conventional structures such as bridges. The protection applies whether the engineering or architectural work is created on paper or on screen or as a model.

Photographs are also protected as artistic copyright works. A photograph is defined as 'a recording of light or other radiation on any medium on which an image is produced or from which an image may by any means be produced'. The intention was to include relatively recent formats, such as holograms. A part of a cinematographic film is excluded from the definition but films have a separate copyright protection (see next section).

2.2.2 Copyright in sound recordings and films

All recordings of sound whether of music, the spoken word or other types of noise, such as a bird song, and any type of image on film, are protected by copyright. The method of recording is irrelevant, so sound on tape, compact disk or digital audio tape are all included. Video tapes and video disks are covered as well as cinematograph films. The intention is that any yet-to-be-invented method of recording sound or moving images will still be covered by copyright. The only requirement is that the record is not a copy of a previous record.

The protection also extends to a recording of a literary, dramatic or musical work made in such a form that sounds can be produced subsequently. This covers making a record by a silent process, for example, synthesised speech created by a computer program.

The copyright in the recording is separate from and additional to other copyrights that may apply. In a recording of a song there will be copyright in the music as a musical work, in the words as a literary work, and in this particular recording. There may be several different sound recording copyrights of the same song. In the case of a film, there may also be copyright in the book on which the script is based and additional copyright in the script itself. Since 1996, a film sound track has been treated as if it is part of the film as far as a copyright is concerned.

2.2.3 Broadcasts and cable programmes

When sounds or pictures are relayed to the public, whether by broadcasts or by cable services, there are additional copyrights.

2.2.3.1 Broadcasts

For radio or television broadcasts, the term used is 'transmission by wireless telegraphy', that is, by use of electromagnetic energy passing through the atmosphere. The broadcast must be intended for reception by the public, although it can be in encrypted form provided decoding equipment is available to the public, for example, in pay-as-you-view satellite services.

Most broadcasts are intended for reception by individuals, but transmissions for limited reception followed by public viewing, such as the live showing at a distant stadium of a sporting event or a pop concert, are also included.

The fact that the signals travel part of the way by cable (e.g. from a TV studio to a transmitter tower; from an aerial by cable to houses in a poor TV reception area; from a community antenna TV system) does not alter the position if the broadcast signals are capable of individual reception elsewhere.

A broadcast has its own copyright provided it is not a repeat of an earlier broadcast. This applies in addition to any copyright in the sounds or images or even a computer program which is being transmitted. For an unscripted situation, such as a sports event, the broadcast copyright may be the only copyright that exists.

2.2.3.2 Cable services

Slightly different provisions apply when the sounds or images are sent electronically, whether by electrical or fibre optic cable or any other path through a material substance. While part of the path may be through the atmosphere, for example, when a live overseas TV programme is provided via satellite, the end user is normally connected by cable, whether the services are available at a fixed time or on demand.

In such an arrangement it is technically possible for information to be sent in both directions. The law is drafted to exclude normal telephone services and confidential two-way services, such as home banking. The wording extends to remote access database services and probably includes home shopping when a viewer receives information on the goods available, although the viewer's actions in placing an order would be outside the copyright provisions. Closed circuit television and TV systems in hotels, hospitals, prisons etc. are excluded from the definition.

The general effect is similar to broadcasting; the transmissions attract their own copyright provided they are not copies of a previous broadcast or cable transmission.

2.2.4 Other copyright works

Copyright also applies to music, plays and conventional artistic works including paintings and sculptures. It extends to works of artistic craftsmanship such as individually designed items of furniture. It applies to typefaces and to typographical editions, that is, the way a literary work is laid out on a page. As there is little of direct application to engineers in these additional types of work, few further references will be made to them.

2.3 Ownership and duration of copyright

In this section, and elsewhere in the book, 'person' can mean either a human being or a company or other corporate body. This is the conventional legal use of the word.

2.3.1 Literary and artistic works

2.3.1.1 Ownership

The creator of a literary or artistic work owns the copyright in it unless this position is varied by a legal contract. One of the most important contracts in this context is a contract of employment.

If an employee creates copyright material as part of his or her job, copyright is owned by the employer. There does not have to be a clause in the contract of employment spelling this out and there does not even have to be a written employment contract, although that is a rare situation these days. It does not matter where or when the copyright work was created. An employer owns copyright in works created at home, in the evening or at weekends. The only criterion is that the work was produced as part of the employee's job.

If an employee creates material quite unconnected with his or her work, then the individual owns the copyright in it. The position is clear when work and hobbies are quite separate and the material in question undoubtedly falls in one area or the other. A production engineer who writes a children's book will own the copyright in it personally. Practical difficulties can arise, however, when work interests and hobbies overlap and there is a grey area. Each case then has to be looked at individually.

Other contracts which affect the ownership of literary or artistic copyright are those which either explicitly set out who is to own it or those which implicitly govern its use because of the surrounding circumstances. When Company A places a contract with an individual consultant or with Company B, for work under which copyright is likely to be generated, the contract terms often include the assignment of copyright to Company A. If the consultant or a representative of Company B signs the contract, then ownership of the copyright is transferred but there must be a document signed by whoever would otherwise own the copyright. The wording can be such that the copyright is transferred in advance of its creation. The effect would be that the consultant or Company B never owns the copyright, it always belongs to Company A. Assignment in advance of the work being done is perfectly permissible. Alternatively, ownership can be assigned after the copyright work is completed.

Suppose, however, the contract makes no mention of ownership: the consultant or Company B will then own the copyright, but may not have full control over it. If Company A pays for the work, then even without gaining ownership of copyright it will have some right to use the copyright material because that was the whole point of placing the contract. If there was a disagreement, a court would almost certainly find that the company paying for the work had the right, by implication, to use it. The extent of the right might, however, be more limited than that given by outright ownership. For example, the implied right to use the material might be limited to Company A and not extend to other companies within the group to which

Company A belongs. Company A certainly could not sell the copyright. Explicit terms are always preferable.

If a commercial argument develops about who is to own the copyright it is often tempting to suggest joint ownership. This is a trap for the unwary. Whereas each joint owner may copy and otherwise use the material internally within the company, granting permission to others to copy or use it requires the consent of both, or all, co-owners. Before agreeing to joint ownership, this severe restriction must be considered very carefully.

2.3.1.2 Duration of copyright

Whether or not the engineer who devises the material owns the copyright in it, he or she is important in determining the duration of the legal protection. Most copyrights last for the life of its creator plus a further 70 years. The 70 year period runs from 1 January following the year during which the creator died. If two or more authors were jointly involved, so that the contribution of one cannot be separated from the contribution of the other, the 70 years begins to run on the death of the last surviving creator.

For copyright works generated before 1 January 1996 the duration is life plus 50 years.

In an engineering environment it is unlikely that copyright will be important for anything remotely approaching this length of time. The life plus 70 year copyright is however important for non-technical books, music, plays and films.

2.3.1.3 Duration of database right

In the case of a database, the right lasts for 15 years from the end of the year the database was created or made accessible to the public. In the case of the IEE's INSPEC database which is continually updated, it is legally arguable that a new copyright is created by every addition, so in theory copyright in it will continue indefinitely. This position is currently under consideration by the European Court of Justice, and at the date of writing it seems possible that this argument will be supported.

2.3.2 Computer-generated copyright

In all examples so far, generation of copyright material has involved a human being. But technology has reached the stage where this is not necessarily the case and UK law caters for a computer itself generating literary or artistic copyright. For example, a computer-generated work is created by an automatically scanning camera, such as those used in security systems or, more picturesquely, by cameras scanning mountain views as seen in resorts in the Alps. Another example of computer-generated copyright is the use of a silicon compiler program to design a semiconductor chip layout as considered in Section 3.6.4. Similarly, if a human writes a program to translate one computer language into another language, when that translation program is run there will be no human author of the translation produced.

If copyright material is created by a computer, copyright is owned by the person making the arrangements necessary for that creation. This seems to imply a financial connection; probably the person or company that owns the computer will own the copyright. The copyright lasts for 50 (not 70) years from the end of the year in which the copyright material was created.

In addition to copyright, computers can also create registered designs, design rights and topography rights (see Chapter 3).

2.3.3 Sound recordings, films, broadcasts and cable programmes

Copyright of a sound recording or film is owned by the person making the arrangements necessary for the recording or the film to be made, conventionally the producer, but since July 1994 the producer and the principal director have been held to be joint authors. In broadcasting, the legal person making the broadcast, for example, a broadcasting company, owns the copyright. For cable programmes the person providing the service, that is, the cable company, owns the copyright.

Copyright in sound recordings or films lasts for 50 years from the year the film or record was made or, if it is released to the public, 50 years from its release. Copyright in a broadcast or cable programme lasts for 50 years from its first transmission. The rights are additional to the separate copyright in a play, song etc. included in the recording or broadcast.

2.4 Marking

Readers will be familiar with the internationally recognised copyright symbol ©, always used on books and often used on other printed matter. The conventional marking contains the copyright symbol, the name of the copyright owner and the year of first publication, as shown at the beginning of this book.

For computer programs, it is highly advisable to mark not only the label on the disk or packaging but also to include a notice on screen, for example, as the first screen display when the program is run. If the available printing fonts do not include ©, then the full word 'Copyright' or an abbreviation, such as 'Copr.', can be used. Databases should also be marked.

While copyright protection in the UK does not depend on there being such a marking, it is advantageous because it reminds others that they need permission to make copies and indicates who to ask for permission. Another good reason for including a mark is that if legal proceedings are started, the marking is presumed to show the true copyright owner until proved incorrect. This means the owner does not need to demonstrate ownership, which can be expensive and time consuming. For sound recordings, films and computer programs, the name is presumed to be that of the copyright owner and any date is also presumed to be the correct date of first publication or showing in public.

In addition to overt marking of copyright, hidden markings can be used, such as made-up addresses in a customer list or redundant code in software. These markings make it easier to show that copying has occurred without permission, especially if the person copying claims to have generated the list or program independently.

The software bugs that are inevitably present in large programs can be used in the same way, and a record of the bug-fixing schedule can even pinpoint which legal version was illicitly copied. See also Rights Management Information in Section 2.7.1.8.

2.5 Moral rights

Moral rights are in addition to the legal rights associated with copyright. There are two major rights which may sometimes be relevant to an engineer, and two minor ones.

Employees only have moral rights in very limited circumstances, so most engineers are excluded, except when they create copyright works completely unconnected with their occupation. However, independent engineers, such as consultants and authors of text books such as this one, will have moral rights as the work is outside their employment.

2.5.1 Paternity right

The first major right, sometimes called the right of paternity, is the right for the creator of a literary, artistic, dramatic or musical copyright work and the director of a film (unless he or she is an employee) to be named as author or director in certain circumstances. These mainly relate to commercial publication of the work and making it available to the public in some way. The name (or initials or pseudonym) must appear on every copy in a reasonably prominent position. The designer of a building or other structure has the right to be named on it in a position that is visible to persons entering or approaching it.

Paternity rights do not apply to computer programs or to any work created by a computer without human intervention (see Section 2.3.2). Material prepared for publication in a newspaper or periodical or for a collective work such as an encyclopaedia, is also excluded, in the last case because keeping track of the multiple moral rights would be impractical.

The right is given only to an individual person and only when there is also a copyright. Paternity right is separate from copyright and cannot be assigned to another person; the only way it can be transferred is in a Will. Copyright and paternity right in the same item can therefore be owned by different people. The right lasts for the same period as copyright, that is, life plus 70 years.

Paternity right must be asserted in writing before it becomes legally enforceable, that is, a statement must be made that the creator intends to benefit from it. Alternatively, it can be waived, also in writing. Readers may have seen moral right assertions in books and published articles. Surprisingly, having the author's name on a published book does not count as an assertion of paternity right; there must still be a separate written statement of the author's rights.

Although paternity rights only came into existence in the UK on 1 August 1989, when the 88 Act came into effect, they are to some extent retrospective. They apply to copyright created before that date, provided the creator was alive on 1 August 1989 and had not assigned or licensed the copyright. The retrospective rights do not apply to films.

2.5.2 *Integrity right*

The second major moral right is integrity right. This is the author's right to object to a distortion or mutilation of a work when it is published, or to changes made which are considered to be prejudicial to the honour or reputation of the creator. In one of the few cases about moral rights reaching a court, it was held that this must mean more than merely the author feeling aggrieved.

The change constituting a distortion etc. must be made to the work itself and not to the way it is treated as a whole, such as including a copyright photograph in an exhibition of which the photographer disapproves. Surprisingly, objections cannot be raised to a normal translation of a book or article, even though a poor translation can give a totally false impression of the original work.

In the case of integrity right, employees are not totally excluded but they only have the right to object to derogatory treatment if they are named either at the time the change is made or on previously published copies of the work. Even this limited right to object does not apply if it is made clear that the creator has not approved of the changes.

The designer of a building has an integrity right but this is limited so that if a derogatory change is made, the creator merely has the right to insist on the removal of his or her name from the structure.

Integrity right does not need to be asserted, but it can be waived. It lasts for the same period as copyright, life plus 70 years. It cannot be assigned, but can be left in a Will.

As with paternity right, integrity right does not apply to computer programs, works created for newspapers or magazines or multi-authored works, such as encyclopaedias, but it does apply to copyright material created before the 88 Act came into force on 1 August 1989, provided the creator was alive on that day and had not assigned or licensed the copyright. There is one exception: films created before that date do not attract integrity rights for their directors.

2.5.3 *False attribution right and privacy right*

The first minor moral right is the right not to be wrongly named, that is, not to have a work falsely attributed to a creator. False attribution right lasts for the life of the creator plus 20 years. The second relates to photographs taken under a commission (i.e. for payment) for private purposes. The photographer owns the copyright but the person paying for the photograph to be taken can stop it being used for other purposes, such as in a newspaper to illustrate a subsequent news item. This is a privacy right, it lasts for the life of the photographer plus 70 years. As with paternity right and integrity right, the owner cannot assign false attribution or privacy rights, but can leave them in a Will.

2.5.4 *Misuse of moral rights*

If any moral right is infringed, for example, if a book is published without naming the author who has asserted the right, or has been altered in an objectionable way, the owner of the moral right can sue for what is known as 'breach of statutory duty', that is, the user has acted in a way contrary to the owner's rights given by the 88 Act. If the moral right owner is successful in court, there can be an award of damages or an injunction to stop the misuse (see Section 9.2.2).

There are some uses of work protected by moral rights which are permitted, such as using it when reporting current events by broadcasts or films etc., or including the work incidentally in broadcasts, films or the like; however, the permitted uses are complex and are not dealt with in detail here.

2.6 Other relevant law

This chapter relates mainly to copyright material that came into being after the Copyright, Designs and Patents Act 1988 came into force on 1 August 1989, but copyright is a very long-lasting right so it is sometimes necessary to apply earlier law to a particular situation. Succinct generalisation is impossible but a few examples will illustrate the need for care.

1. The previous Copyright Act of 1956 came into force on 1 June 1957. Before that date, broadcasts were not copyright, and the sound track of a film was considered to be separate from the film; it was a sound recording.
2. Under the 88 Act, copyright in a photograph belongs to the photographer, whereas under the previous act, copyright in a photograph belonged to the person owning the unexposed film.
3. It is not only the previous copyright act that must be considered as other laws may be relevant. For example, cable programmes did not attract copyright until the Cable and Broadcasting Act 1984 came into force on 1 January 1985.

For any copyright work created before 1 August 1989, sources other than this book should be consulted.

2.7 Copyright use and misuse

Ownership of copyright gives two types of legal control. The first, as the name implies, is the right to stop others from copying the material without permission: this is known as primary infringement and is covered in Section 2.7.1. The other type is the right to stop the trading of those illegal copies, called secondary infringement (see Section 2.7.2).

What the owner does not have is the right to control the idea behind the copyright material. Copyright law applies only to the form in which an idea is expressed, to the actual words or the actual drawing, not to the concepts on which the words or drawings are based. So if a technical report describes a test method, it is not infringement of copyright to carry out the test (but remember it may be patented). You may not copy the words of the description, but you can describe the same test using different words. If a computer game is based on a new idea, it is permissible to devise another game using the same idea, but not to copy the code or the graphics.

2.7.1 Infringement by copying – primary infringement

2.7.1.1 Copying in general

If someone copies without the copyright owner's permission, that person is liable to legal action unless the copying falls within a variety of exceptions that are summarised in Section 2.8. If a copy is made, there is infringement. It is no excuse that something was added. It is no excuse that you did not do the copying yourself; authorising someone, whether an employee or a third party working under contract, to make an illicit copy is still infringement.

The copying need not have taken place from the original copyright work. Copyright in this book is based on a word-processed text, but that copyright is infringed by copying the printed version. It is even possible that copyright in a drawing could be infringed by working from a verbal description of it.

However, there must have been an act of copying. Two sets of mathematical tables will be indistinguishable, but if each devisor can prove it to be an independent creation there has been no copying and no infringement. If a photographer takes a photograph of a particular scenic view, it is not infringement for another person to photograph the same view, even though the two photographs are virtually indistinguishable; the second photograph has been independently created. It would, however, be infringement to create a scene in a studio to reproduce a scene in someone else's photograph: in that case, there is artistic copyright because the scene is set up artificially and taking a photograph of the recreated scene is an infringement.

A strong similarity is not always proof of copying. One also needs to show that the alleged copyist has had access to the material which has supposedly been copied. On the other hand, if there is a similarity and access to the original was possible, one may infer that the material has been copied.

2.7.1.2 Copying in a different medium

If a copy is made in a different medium to the original work, such as a photocopy of a book, or a scanned-in digital copy of a photograph, it is still an infringement. So long as there has been a reproduction and the copy has a material form, the law applies to it. So far as modern technology is concerned, there is a specific reference to 'storing the work in any medium by electronic means', so whatever the format, a copy has been made.

Making a short-term or transient copy is still infringement, to cover use of volatile memory. Putting words onto a computer screen then deleting them is still a form of copying. The reader can derive considerable benefit in viewing, say, the contents of a database, so the definition is an essential one, giving rights of control to the provider of a database service.

This is reflected in the charging mechanism for online access to commercially available databases, such as IEE's INSPEC technical abstract service. There are three main ways of access: (a) password-controlled access online where the customer is charged connection time plus a record displayed charge; (b) password-controlled access via a web service offering pay-as-you-go connection without a connection charge but with a per record displayed charge and (c) by use of an annual site licence where a fixed price gives unlimited access during the year.

An example of an immaterial form, in which case there is no infringement, would be speaking words aloud without making a copy whether by shorthand or on tape. This type of use is not copyright infringement, but depending on the circumstances it could count as a performance when it will be covered by the 88 Act in other ways. A performance is not necessarily a theatrical performance, and can include delivering a lecture or a speech, so repeating someone else's lecture needs permission.

For artistic works, it is infringement of the copyright to make an object in three dimensions from a drawing of that object, but for engineering drawings the special provisions of design right (Section 3.2) apply.

For architects' plans of buildings and fixed structures, it is infringement of the plans to construct the building as well as to copy the plans themselves. Conversely, copying a building in two dimensions, such as taking a photograph or making a sketch of it, is not an infringement if the building is on public display, by a special exception.

2.7.1.3 Copying part of a work

It is still infringement of copyright if one does not copy everything but uses a 'substantial part'. This can be judged by quality as well as quantity. In a non-technical context, one line of a well-known song would be considered a substantial part. Copying one frame from a film infringes copyright in the whole film. In general, if one is benefiting from use of the copyright work by saving oneself time and effort instead of producing an independent version, then usually there has been infringement.

2.7.1.4 Copying by adaptation

It is primary infringement of copyright to adapt a copyright work, which includes translating it. This can be conventional translation of a book from English into a foreign language, but the infringement occurs as soon as the translation is recorded in any way whether on paper or disk.

2.7.1.5 Copying computer programs

Computer programs get special attention in the 88 Act. While the word 'translation' is usually associated with human languages, the wording of the law makes sure it is applied to computer languages also. The owner of copyright in a program can control its conversion into a different computer language and its conversion into or out of computer code. Conversion from source code to object code is therefore included, and vice versa.

Computer programs are necessarily a special case as far as copyright law is concerned because their normal use inevitably involves an act of copying. When an applications program on a disk is loaded onto a personal computer (PC), a copy is made in a different medium. When hardware is running software, parts of the software code are copied internally, for example, from a read-only memory into an instruction register; in other words, a copy is made of a part of the program. The copy may be transient, but making it is still an infringing act.

Because computer programs cannot be used without an act of copying, it is legally essential for the copyright owner to give permission for the copying. This is usually done in a licence specific to the software, especially for applications programs. For operating software (i.e. the software that controls the internal functioning of hardware and how it interacts with peripheral equipment), there is almost certainly an implied right. The hardware cannot be used without the software, so supply of the hardware implies the right to use it. Sometimes an operating software licence is granted specifically. For programs on semiconductor chips, for example, in cars and washing machines, there is an implied right to use.

2.7.1.6 Copying databases

The extraction or reutilisation of a computer-stored database, whether the whole or a substantial part, is an infringement of database right. There is no exception for fair dealing, that is, copying a small part only. This is reasonable because most use of a database involves scanning the contents and rejecting most of it, with only a very small part being positively selected and stored by a legal user.

2.7.1.7 Broadcasting and other public use

A TV, radio or cable programme has its own copyright, so it is infringement of copyright to re-broadcast it or include it in another cable programme. It is also infringement of other types of copyright works to include them in broadcasts or transmit them by cable whether the work is a song, a film, a drawing or any other type of copyright material. Transmission by telecommunications system is also included and the words apply to short-distance transmission, for example, in a computer network.

A copyright owner also has the right to stop anyone issuing copies to the public, provided the owner has not already done so. Therefore, even if one legitimately purchases copies of a photograph or a report, one is not necessarily permitted to put them into public circulation. If, however, the owner has issued copies to the public, one can then treat the legally acquired copies in the same way as any other property and they can be sold or hired out. This applies unless the copy is a sound recording, film or computer program, in which case only the owner is allowed to rent them to the public. The reason is that with modern technology, perfect copies of such recordings can be made very easily so the owner needs extra control.

2.7.1.8 Copy protection

Because computer programs are so easy to copy, some software houses incorporate built-in copy protection devices. Devising a program to disable such a device counts as infringement of copyright in the protected program.

Some owners build a Rights Management Information system into Internet products using, for example, digital watermarks. The information can identify the work, its author, its owner and the terms and conditions of use of the work.

The owner can then keep track of the use of the product and can enforce copyright if necessary. Removing such a system is not allowed and nor is trading in devices to remove it.

2.7.2 Infringement by trading – secondary infringement

2.7.2.1 Guilty knowledge

Even if a person does not actually do or authorise the copying without the owner's permission, a variety of ways of dealing with the work commercially can still be infringement. In all cases the person must know, or have reason to believe, that the copies have been made without permission. The phrase 'have reason to believe' extends a long way. The courts will assume that engineers have a general knowledge of the laws applying to their work, including copyright law. This would be especially true if they dealt with copyright material, such as computer programs, on a regular basis. A claim of sheer ignorance of the 88 Act would not necessarily give an escape route, unless copyright law was extremely remote from one's normal work. Readers of this book are, of course, now generally aware of the law.

2.7.2.2 Imports from abroad

Acquiring copies abroad and importing them into the UK is secondary infringement of copyright. The only exception is for copies imported for private and domestic use. This is hardly likely to extend beyond one copy, which should not be used for any business purposes.

So buying a single copy of a computer program in the Far East, where software pirates flourish, and bringing it home, is allowed under UK law, but not for use in connection with professional engineering work.

The right to stop imports may also apply to legally made copies. The test is whether making a copy in the UK would be infringement of the rights of the copyright owner or licensee. There could be infringement of commercial rights. For example, books are often sold in different countries at different prices; import into the UK of low-priced books would be objectionable to a company selling books here at a higher price. The import can sometimes be stopped, depending on the origin. EU regulations allow free movement of goods once they are on the market, so there is no control over books originating in another EU country. A low-priced import from outside the EU could, however, be stopped by the owner.

If a copyright owner finds out that import of multiple copies is planned, the customs and excise authorities can be approached and a request made to treat the infringing copies as prohibited goods. Obviously the more information provided about the shipment, the better the chance of it being identified. Suspect imports are held in bond until they are checked by the copyright owner.

2.7.2.3 Business use

If an engineer knows, or would be assumed by a judge to know, that a copy is illicit, almost anything the engineer does with it could lead to legal action. The copyright owner can sue for infringement if a copy is sold or hired, or offered for sale or hire. The same applies if the article is distributed either internally within a business or outside the company. Even possessing the infringing copy in the course of a business is an infringement. In other words, doing anything which has a professional or commercial link can be an infringing use.

Transmitting an illicit copy over a telecommunications network is infringement if the person controlling the transmission knows that an infringing copy will be made when the transmission is received. The wording is such that short-distance transmission, as in computer networking, can be included.

The law extends even beyond the commercial dealings with illicit copies mentioned above. It is also infringement to produce an article which is designed to make an illegal copy of a particular copyright work. An example in the music industry would be making a master from which compact disks can be reproduced without permission. This would also apply to importing the master copy, selling it, hiring it or possessing it in the course of business.

General copying devices, such as photocopiers, are not caught by the provision because they are not linked to one specific copyright work.

2.7.2.4 Public performances

Engineers whose normal employment involves supplying apparatus for playing sound recordings, showing films or receiving broadcasts or cable programmes in public can leave the question of copyright permission to their employers. But 'public' can mean an audience of parents at a school, or the general public in a village hall, and can include members and guests in a club. The performance of a school play watched by parents requires a licence from the Performing Rights Society, because performing a play is otherwise infringement of copyright. Similarly, showing a TV broadcast, cable programme, film or playing music to the public requires permission.

Even if the reader is not involved in the primary infringement, for example, acquiring a copy of the film, there are other types of involvement that could be a cause for concern. One is supplying the apparatus (or a substantial part of it) on which the film or broadcast is to be shown – so lend your video equipment with care! The other is allowing the hall to be used for the event – school governors and parish councillors take note. Both are secondary infringements of copyright if the person supplying the apparatus or authorising use of the premises realised that showing the film etc. would be primary infringement of copyright.

There is one exception, which is playing sound recordings in a club, but the definition of a club in this context is limited.

2.8 Exceptions to copyright infringement – fair dealing

There are several exceptions to the control exercised by a copyright owner. Those of interest to engineers are mainly the exceptions known generally as 'fair dealing'. In this context the word 'dealing' does not imply any commercial transaction or even the involvement of more than one person, but simply means using the copyright material in some way.

2.8.1 General business

It is permissible to copy a literary, artistic, musical or dramatic work if one needs the copy for research or private study. The research must be for non-commercial purposes so all business-related acts fall outside the exception. It does not mean that a librarian at a company research laboratory is allowed to make copies for internal circulation. The permissible amount copied is limited to one complete chapter of a book or up to 5 per cent of a literary work, by agreement with societies representing authors.

Librarians in certain designated libraries, such as the IEE library, and educational and public libraries, can make the single permitted copy for you, but are subject to severe restrictions on making multiple copies of the same article or extract. The designated libraries are also required to charge a fee for the copying, and usually there will be a form to be signed stating that the copy is being made for fair dealing purposes.

It is also fair dealing to quote someone else's words or equations for the purpose of criticising or reviewing the work in question. There must be sufficient acknowledgement, that is, it must be made clear that a quotation is being used and its source must be identified.

As far as computer programs are concerned, lawful users may make a back-up copy for security but it is not fair dealing to decompile software in general even for research purposes. A licensee may decompile to permit the creation of a compatible program provided the information needed is not otherwise available and licence agreements must not forbid such decompilation.

2.8.2 Public events

It is also fair to include copyright material when reporting current events, provided there is sufficient acknowledgement of the original work. Any type of copyright work can be used in this way, but if the work is reported by a sound recording, film, broadcast or cable programme, there does not have to be an acknowledgement.

Taking an example of an actual case, when the football World Cup competition took place in June 1990, the BBC had monopoly rights on televising the matches in the UK. British Satellite Broadcasting Ltd (BSB) used short excerpts, less than a minute, of match highlights in its sports news programmes. The fact that the material originated with the BBC was acknowledged, even though this was not a requirement. The BBC sued for copyright infringement, but in court the use by BSB was held to be fair dealing.

If an artistic work, recording, film etc. is being made and another copyright work is included as part of the background, then such incidental use is not infringement. If a broadcast commentary of a football match picks up copyright music being played over the loudspeakers at the ground, or if a film of a street scene includes a pictorial trade mark which is a copyright work, then no infringement has occurred by reason of this exception. But if the music or the trade mark was deliberately included as a feature of a programme, it would be infringement, not fair dealing.

Exceptions to copyright control also apply to buildings, structures and models on public display. There is no control over anyone making a drawing, taking a photograph or including the objects in a TV broadcast or cable programme, but a photograph of a current event may not be used without the permission of the owner of copyright in it.

2.8.3 *Time shifting*

A special exception is made for recording broadcasts or cable programmes for the purpose of time shifting for more convenient viewing in domestic premises. There is no time limit on how long the copy can be kept.

It is also permissible to take a still photograph of a TV broadcast or a cable programme, again for private and domestic purposes only.

2.8.4 *Education*

Schools, colleges and universities have greater freedom to copy than individuals or companies, provided the copying is done for instruction or for setting exams.

The act does permit schools to make a limited number of photocopies, which is broadly speaking 1 per cent of any copyright work in any quarter year. This does not include sheet music, maps or newspapers, and does not apply if there is a licensing system in place.

Such a licensing system is run for UK universities, polytechnics and schools by the Copyright Licensing Agency Limited (www.cla.co.uk), and allows greater copying than the exception in the act. Payments are made by a central authority to the copyright holders. In local authority schools, a sampling system is used and the photocopying results grossed up to calculate the payments, which cover over 200 million photocopies each year. Schools in the private sector also have an agreement. In all cases the permitted copying is limited to use for study or teaching.

As far as sound recordings, films, broadcasts and cable programmes are concerned, educational establishments are permitted to make copies, provided this is done by the teacher and for instructional purposes only.

2.8.5 *Technical abstracts*

If a technical paper is published in a periodical with an abstract, then copying the abstract is permitted, unless there is a licensing scheme in force. This allows the IEE's abstracting service INSPEC to use author-written abstracts without the need for permission from each individual.

2.8.6 *IEE papers*

It is a requirement of the IEE that it owns the copyright in the articles published in its journals, with rare exceptions such as Crown copyright. All IEE journals contain an explicit copyright notice:

> © 2005 The Institution of Electrical Engineers. All rights reserved. No part of this publication may be reproduced, stored in a retrieval system or transmitted in any form or by any means without the permission in writing of the publisher. Copying of articles is not permitted except for personal and internal use, or under the terms of a licence issued by a Reproduction Rights Organisation such as the Copyright Clearance Centre Inc (Ref: 0960-7919 $20.00). Multiple copying of the content of this publication without permission is always illegal.

The institution is a member of the licence schemes applying in schools and universities, as mentioned above.

2.8.7 Central licensing scheme

For many years there has been a central body dealing with licensed copying of books and journals on behalf of authors and publishers. The Copyright Licensing Agency (www.cla.co.uk/licensing) authorises copying, collects royalty payments and passes the appropriate percentage to the author or publisher. The agency also administers two general photocopying licence schemes, one for companies employing up to 50 people and another for larger companies when the licence fee depends on the number of professional employees in the firm.

2.9 Taking action against infringers

If the owner of copyright believes the material it protects is being copied without permission, an action for infringement can be started in the civil courts (see Chapter 9). If the owner wins the action, the court may impose three possible penalties:

- an injunction to stop further copying
- payment of damages to recompense the owner
- delivery-up of the infringing copies.

Any combination of the three remedies may be applied.

The owner will need to be certain of ownership of the copyright, and good records showing which employee or employees created the work are helpful, or a clear assignment document transferring the rights from a third party. The owner also needs to be able to show that the allegedly copied work was accessible to the person accused of copying. An exclusive licensee has the same right to sue as the owner.

The alleged infringer can argue in defence that he or she did not know the work was copyright. This does not imply that an engineer can argue ignorance of copyright law, it is directed more towards belief that copyright had expired or did not extend to this type of material.

For the type of copyright material likely to be sold on a market stall, such as computer games, there is a special right. The copyright owner may seize the copies in person. The right is conditional: the goods must be on display for sale or hire; force must not be used; and the local police must be informed in advance. The right does not extend to copyright material displayed in a permanent place of business, so the possibilities are limited.

2.10 Criminal liability

2.10.1 General criminal offences

Some types of copying or of trading in copyright material are criminal offences. The scope is narrower than the range of infringements set out above, is more closely linked to business purposes, and stricter proof of the facts is needed. In many cases there must be an 'article', in the normal sense of a three-dimensional object, and the person committing the offence must have known, or had reason to believe, that the article was an illegal copy. This person will be assumed by the court to have a general knowledge of copyright law, so the defence of ignorance is unlikely to be available in most cases. The criminal provisions also extend to public performances of literary, dramatic or musical works and publicly showing a film or playing a sound recording.

The criminal offences relating to an article that is known to be an infringing copy are:

1. making the article for sale or hire
2. importing the article into the UK (except for private or domestic use)
3. possessing the article in the course of a business with a view to committing any act infringing the copyright (e.g. intending to copy it)
4. selling, hiring, exhibiting or distributing the article in the course of a business
5. distributing the article so as to affect the copyright owner's rights in a prejudicial way (e.g. depriving the owner of legitimate sales or licence fees).

The penalties are up to 10 years imprisonment (an increase from 2 years was made in 2002), an unlimited fine or both. In the case of a company found to be offending in this way, company officials, for example, a director, the company secretary or senior manager, may be held personally liable if they are aware of what is occurring.

The criminal activities most likely to be encountered by engineers relate to computer software, but similar principles will apply to other types of copyright work.

2.10.2 Criminal misuse of software in business

Making illicit copies of computer programs is extremely easy; all one needs is appropriate hardware, and in many cases a PC is sufficient to give perfect reproduction. There are two main types of misuse. The first is making copies for supply to third parties at a lower price than the cost of a genuine licence; this is often called software piracy. The other is copying without permission within a company; this will be referred to as corporate copying or underlicensing.

If a company pays for one copy of a program for a PC, such as a word processing or accountancy package, it does not have the right to make copies for every PC in the company, unless the accompanying licence gives this permission. Usually it will not, and further licences in return for payment are required. This section explains how far the criminal provisions apply to copies made without such a licence. Since this type of copying is widespread, the legal wording will be considered in detail.

The 88 Act states:

> A person commits an offence who, without the licence of the copyright owner . . . in the course of a business distributes, or distributes otherwise than in the course of a business to such an extent as to affect prejudicially the owner of the copyright . . . an article which is, and which he knows or has reason to believe is, an infringing copy of a copyright work.

An 'offence' means a criminal act.

If copies of a program on floppy disk are made, the copies are 'articles' in the normal sense of the word, that is, three-dimensional objects. If company personnel make copies for internal use within the company, this is highly likely to be 'in the course of a business'. The word 'distributes' can be applied to supplying the copies internally within the company.

If legal interpretation did not regard this internal distribution as being in the course of a business, then the phrase 'distribution otherwise than in the course of a business' is applicable. One alternative or the other must apply. The copying and internal supply are certainly prejudicial to the owner of the copyright, who could otherwise have expected licence fees for the copies.

The phrase 'he knows, or has reason to believe' has been interpreted by the Australian courts in the context of the Australian Copyright Act, and the Judge said 'knowledge cannot mean any more than notice of facts such as will suggest to a reasonable man that a breach of copyright law was being committed'. This statement has been approved in cases in the English courts. Probably, a very junior employee copying on his or her own initiative would be able to plead ignorance as a defence, but the 88 Act gives more detail on responsibility as follows:

> Where an offence . . . committed by a body corporate is proved to have been committed with the consent or connivance of a director, manager, secretary or other similar officer of the body, or a person purporting to act in any such capacity, he as well as the body corporate is guilty of the offence and liable to be proceeded against and punished accordingly.

Wording similar to this is found in many places in English law. In one case it was held that where the director consents he is 'well aware of what is going on and agrees to it', and that where a director connives he is 'equally aware of what is going on but the agreement is tacit'. The company director etc. would therefore need to have specific knowledge of the copying of the disks. From a practical viewpoint, a director of a small company is much more likely to have knowledge of such activities than a board member of a large corporation.

The level of liability has also been considered in a different legal context. It will extend only to very senior executives, possibly only those who are named in the company accounts. It is unlikely to apply, for example, to a purchasing manager in a large company who knew that programs were always acquired as single copies.

At the date of writing, there has been no full hearing in court on corporate copying, but the interpretation above is based on an opinion by a senior barrister and it has been sufficiently persuasive for several large organisations to settle out of court.

2.10.3 Federation against Software Theft (FAST)

Software misuse, whether within a company or for outside sale, causes the industry enormous losses. Currently, it is believed that on a global basis one-third of software is pirated. Even in the early 1980s the problem was clear, and a number of interested companies collectively set up FAST in 1984 to lobby parliament to change copyright law.

FAST currently has about 160 members from the software publishing industry including resellers, distributors, audit software providers and consultants. FAST and its members recognise that controlling the use of software within a company is extremely difficult, and one of its main aims is to assist managers in identifying and rectifying any software misuse by their staff. To this end, FAST Corporate Services Limited has a software compliance programme called the FAST Standard for Software Compliance (FSSC) which is recognised by the BSI and takes members through a three-stage programme to reach 'Registration'. FAST Consultancy Services Limited provides consultancy services under which a company's IT asset base can be reviewed allowing software compliance to be achieved and IT investment optimised. Alternatively, FAST Consultancy Services Limited is the UK distributor for Software Organiser, a tool that allows a business to detect and list all its software applications including those in Personal Digital Assistants; this provides a single reference point for ongoing management of all software licences and helps with internal compliance.

Proper control of software significantly reduces the risk of infringement actions and also from other risks, such as infection by computer viruses. Such viruses are rarely found in software provided by an authorised source. A similar situation exists in the USA where the piracy rate is 24 per cent and fines for software violations exceeded US$13 million in 2001.

For flagrant criminal copying of software, FAST helps its member companies to initiate criminal prosecutions. Success has been obtained in the UK against both corporate copying and software pirates. FAST is contactable at www.fast.org.uk., by e-mail at fast@fast.org or by telephone on 01 628 622 121.

2.10.4 Other types of copyright piracy

Several types of copyright material other than software offer the possibility of making profits from illegal copying. The obvious examples are recordings of popular music and videos. Sometimes, the copying technology is only available in a business context, sometimes it is available at home. Piracy of such material takes place on an international basis, and in all countries.

Historically, the first problem area was probably in the popular music field when illegal pressings of disks were made. The record companies rights are looked after in the UK by the British Phonographic Institute (BPI) which assists members in bringing civil and commercial cases to court, and internationally through the International Federation of Producers of Phonographic and Videogames (IFPI), which lobbies for improvements in the law in countries where copying is most flagrant.

When video recorders became widely available for home use, it was the policy of film companies not to issue video copies of their expensive films until a year or so after release to cinemas, so that profits could first be made in this area. A back street trade quickly developed, with films being 'borrowed' overnight and copied onto video tapes. The technology allowed copies to be made quickly by using professional equipment. In the UK, the film companies retaliated by setting up the Federation Against Copyright Theft (FACT) with a team of investigators who collected evidence for criminal prosecution of those responsible for illegal copying. FACT succeeded in markedly reducing the extent of such copying by making it a high-risk business, but it still occurs and the organisation is still active, see www.fact-uk.org.uk.

Another current area of major piracy lies in books. For high-quality copies, commercial equipment is needed and illegal copies often originate in Far East countries.

2.11 Copyright licences

A copyright owner has very wide legal rights to stop copying, so permission is often needed before a copyright work can be used. Reading a book does not require a licence because there is no act of copying, but an author, illustrator or photographer will need to grant their publisher the right to copy their material before the book can be printed. This permission is usually granted in a licence, a written agreement setting out to what extent the copyright work can be copied, and the payments to be made in return.

Licence agreements are almost infinitely variable in content, and the general principles are covered in Chapter 9. The licence may grant the right to copy manufacturing drawings to prepare for a major production contract, but the most common type of copyright licence is for use of computer software.

If a major software suite costing many thousands of pounds is acquired, the contract documents to be signed will include a licence to run the software. For software bought over the counter, the package including the disk and user manual very frequently contains a short licence agreement; often this is visible through the plastic wrapping and includes a statement that tearing open the wrapping is assumed to mean consent to the licence terms. These are called 'shrink-wrap licences'. For software downloaded from the Internet a 'click-wrap' licence is used; before the download the licence terms are displayed and the potential user is asked to click an 'I agree' box. The legal principles involved in shrink-wrap and click-wrap licences are far from sound, but the technique is widely used as it is the only possible way to impose contractual obligations in this type of commercial deal where the purchaser does not sign a document.

In all cases, the copyright owner has the same intention, to make it clear that the user can only copy to a limited extent, such as running the software on one PC at a time. For more powerful hardware, the right may be limited to use on equipment bearing a particular serial number. If the licence does not say that networking is permitted, then it is not. Usually, wider rights are available for further payments.

Unfortunately, there is no standard software licence and no standard wording. For each program the relevant agreement must be consulted, and the wording of some licences leaves a great deal to be desired so far as clarity is concerned.

2.12 Copyright internationally – general and non-technical works

2.12.1 The Berne Convention

The Copyright, Designs and Patents Act 1988 applies to the UK only; copyright laws are always national laws. Import and export of books has taken place for centuries, and for decades there have been international agreements on copyright, making the protection automatic to a large extent in many countries. Most developed countries belong to the main convention, the Berne Convention dated 1886, which in September 2004 had 158 member countries, the major exceptions being some countries in the Middle and Far East. The USA joined in 1989 and China joined in 1992.

The general arrangement is that member countries give the same legal protection to copyright material originating outside that country as applies to copyright generated within the country. Copyright should apply without any registration or other formalities. There is no obligation to add a copyright marking and the minimum protection is for life of the author plus 50 years.

Under English law, very little creativity is required for a work to be copyright; so long as it has not been copied, it is legally protected. In contrast, Germany requires a higher level of creativity than other countries, so many works will be unprotected there which would have been copyright elsewhere. Other European countries have stronger requirements than the UK but weaker than Germany, for example, giving protection only if the author's personality is imprinted on it.

Some countries operate a home taping levy, that is, they apply a tax on blank tapes for sound and/or video recordings. Germany also taxes recording equipment.

One principle overturned in the USA was the 'sweat of the brow' test. If a not-very-original work had involved a great deal of effort, then that used to be sufficient to attract copyright protection. In 1991, a decision was made that a telephone directory was not copyright even though compiling it had required considerable labour. Significant originality is now required.

In the USA, there is the possibility of registering copyright, although there is some protection without registration. Registration is essential when suing for copyright infringement. Marking with a copyright symbol is also advisable, as it means that the company sued cannot claim to have infringed innocently, when substantially lower damages would be payable.

Readers may have noticed in books, usually paperbacks, the words 'Not for sale in the USA'. The reason is that the USA requires books sold in the country to be typeset there (or in Canada since 1989), although a small number of books can also be imported.

It is also possible to register copyright material in Canada and in Spain. The few examples above should be sufficient to illustrate the dangers of assuming copyright law is universal despite the international agreement.

2.12.2 TRIPS – Trade-Related aspects of Intellectual Property Rights

The TRIPS Agreement places on countries which have signed it the obligation to comply with the Berne Convention, and clarifies some points. TRIPS confirms that copyright protection is to extend to expressions, but not to ideas or procedures or mathematical concepts. It also confirms that computer programs, in both source code and object code, are to be protected as literary works for at least 50 years. Databases are also eligible for copyright protection, even if they exist only in machine readable form.

For most types of work, copyright is to last for at least the life of the author plus 50 years, although for some works including photographs a shorter term may apply. Readers are reminded that different countries will phase in the arrangements at different speeds.

2.12.3 WIPO Copyright Treaty

By the late 1980s, the ease with which digitally recorded copyright works can be copied perfectly, and illegally, was clear and in 1996 the WIPO Copyright Treaty was adopted to deal with relevant aspects of technical copyright plus some other areas.

The treaty restates that copyright extends to the expression of an idea but not the idea itself. It covers databases, that is, compilations of data which by reason of the selection or arrangement constitute an intellectual creation, but the EU regulations on databases predate the treaty and give wider rights. Member countries must provide adequate legal protection against circumvention of copy protection devices – in the UK these were put in place by the 88 Act, see Section 2.7.1.8 – and for rights management systems. Legal enforcement systems must also be put in place.

The treaty extends the rights of authors. For computer programs, cinematographic works and phonograms (sound recordings) the author is given the exclusive right to authorise commercial rental to the public. For the broader category of literary and artistic works, authors are given the exclusive rights to authorise communication to the public, whether by wire or wireless means, by cable or the Internet.

There is also a WIPO Treaty on Performances and Phonograms which gives performers authorisation rights and some moral rights.

2.13 Technical copyright

2.13.1 Various directives in Europe

Many of the provisions of English law relating to computer software and databases have been made to comply with EU directives. The aim is to harmonise the law in all EU countries. Minimum standards are set and individual countries may set higher standards if they wish. With (currently) 25 members, national laws are likely to vary for quite some time.

2.13.2 Technical copyright in the USA

In the USA, copyright law applied to technical areas, especially computer software and use of the Internet, is developing rapidly as decisions are made in major copyright lawsuits. Copyright has been extended to cover the 'look and feel' of a program, that is, the way the human user interacts with the program, even if the code itself is not copied from earlier well-known software. 'Look and feel' is not a defined term, and includes the appearance of screen displays, how they are laid out, and the command words or abbreviations used for common functions. If a user is familiar with one well-known program, the aim of copying look and feel is that another program should be easy to learn because the appearance is familiar. Recent decisions on copyright protection for screen displays are making it more risky to copy their appearance.

Also in the USA the test for copyright infringement so far as programs are concerned used to be based on the structure, sequence and organisation of the code, as well as on line-by-line comparisons. In 1992 this test was rejected and a new 'abstractions' test developed, in which generalisations are made from object code through source code to the ultimate function of the program. The test is to search for patterns; if the patterns are close, infringement is likely. This is referred to as non-literal copyright – most copyright infringements look at literal copying.

The USA naturally provides the earliest examples of the possibilities of flouting copyright law and a well-known case relates to popular music. In 1987 the Motion Pictures Expert Group set up a standard known as MP3 for storing music in compressed form. Napster made its 'MusicShare' software available without charge so that Internet users could (1) index works stored on hard disks in MP3 so that searching and transfer was easy; (2) search the indices of other users and (3) connect PCs together and transfer exact copies. Napster did not itself copy or store the files but in July 2000 a US court found the company liable for vicarious or contributory infringement in a peer-to-peer setting, and a preliminary injunction was granted to stop this. Napster was forced to start policing its system and to delete files where the copyright owner objected, which was the great majority.

Another example relates to digital versatile disks, DVDs, which can store full-length motion pictures. In 1997 eight major US producers began to issue DVDs, but with the obvious risk of illegal and highly profitable copying, they used an access control and copy prevention system, Content Scramble System (CSS). To decrypt and play, a user needed a licence which was available with Windows packages. A young Norwegian who was a Linux enthusiast (see Section 2.14), and did not approve of Microsoft or of Windows, reverse-engineered a licensed player, then created a decryption program to run on Linux. As is the common practice with Linux users, he put it on the Internet for use and comment so knowledge of it became widespread.

This contravened the US Digital Millenium Copyright Act (DMCA) which came into force in 1998 and implements the WIPO Treaty. Under the DMCA it is illegal to circumvent copy protection devices, but what action can one take against a 15-year-old resident in another continent?

The DMCA also allows Internet service providers (ISPs) to store data on servers, to direct users to third-party data, to make transient copies and to put into a cache material which is frequently requested – that is, the usual ISP activities are allowed and are not infringements of copyright.

2.14 Copyleft

The word is generally used to mean 'an agreement which allows software to be used, modified and redistributed freely on condition that a notice to this effect is included with it'. Copyleft licences are monitored by the Free Software Foundation to ensure that these conditions are kept. The concept originates with users of the Linux operating system who number over 10 million worldwide. Their argument is that a code which is accessible to everyone will be debugged very quickly and will be less liable to crash. It is a condition of the payment-free use that if a user improves a program, it must be made freely available to everyone else.

Linux-based code is widely used on the Internet, and some large companies are softening their attitude to free licensing for some products.

2.15 Managing copyright

Given a general knowledge of copyright law, which this book should provide, normal management practice should aim for proper control of copyright material, both in the sense of protecting an asset of the engineering company owning it, and of avoiding legal action for infringement of third-party copyright.

Accurate record-keeping is the first requirement. A note of which employee or team creates copyright material will allow proof of ownership when it lies automatically with an employer. The conventional drawing office system of putting a draughtsperson's name on every sheet of paper and of recording amendments to that sheet is an excellent practice, but not easy to impose in other areas where pre-printed paper is not the norm.

Creating full records can be quite a task. A major software program may be created by one team of software engineers, another may debug the program and a third update it: all contributions need to be known if ownership is ever put to full legal proof.

It is also essential to remember that some types of article have more than one type of copyright. A product brochure or a book may contain text, drawings and photographs, that is, both literary and artistic copyright.

The use of temporary staff such as software engineers or draughtspersons from an agency may cause ownership difficulties. Often the only written record is an invoice; this will not transfer copyright in the program or the drawing from the individual to the company. To do that, a document signed by the copyright owner is needed. Usually this will be the individual engineer, because agencies are rarely employers of the staff they provide.

The right to sue for copyright infringement may also be based on an exclusive licence, which gives the licensee the same rights as the owner, so knowledge of this type of agreement may be important to a manager. The simple rule is that material under third-party copyright should only be copied if the copying either falls within one of the fair dealing provisions set out in Section 2.8, or is permitted in a licence agreement. The licence will record what copying is allowed, and familiarity with its terms will allow potential problems to be avoided.

One clear factor in copyright infringement is that copying is almost always done knowingly. There was a court case some years ago relating to unconscious copying of a song by another composer, but any copying by an engineer is almost certain to be a conscious act. The knowledge that to copy without permission is to risk an infringement action is something which all engineers need to bear in mind.

Copyright can be a major asset with a high degree of awareness among engineers and proper control, or it can be a liability, if copying without permission occurs.

2.16 Summary of copyright

This broad right applies widely in the engineering field, covering reports and specifications, engineering drawings and computer software. Protection is automatic, no action is needed and the legal right lasts for the life of the author/draughtsperson/ software engineer plus 70 years.

The main practical need is for good records of who created the copyright material. It is important to know whether the creator was an employee, in which case the employer owns the copyright, and whether there was a contract with another company relating to copyright ownership.

Marking with the copyright symbol ©, the owner's name and the date is not essential but is a useful warning to others.

The owner of copyright can control copying of the material by others, but not the use of the information it is based on. For example, if an engineering specification is rewritten in different words, there is no copyright infringement.

To avoid infringement, engineers should copy words, drawings or computer code only if they or their employer own the copyright, or there is a clear right to copy which is recorded in a licence agreement.

Chapter 3
Rights in designs

3.1 Introduction

To the general public, the word 'design' probably means the external appearance of an object, its visual appeal. To an engineer, the word is more closely linked to the way the object functions, but good engineering design is also aesthetically satisfying.

One of the definitions of 'design' in the *Shorter Oxford Dictionary* is 'adaptation of means to ends', which applies to both the artistic and the practical design processes. In each, an inherent skill in the creator is developed by training, then used innovatively, whether on paper, on screen or directly on materials.

In the UK, the benefit of innovative design to the economy is well recognised, and both artistic and functional designs are legally protected. Just as one type of design merges into the other, so do the applicable areas of law, which are copyright, design right and registered design.

Chapter 2 explains how copyright applies to drawings, whether intended to be artistic or produced in an industrial context. This chapter looks at the legal position when a three-dimensional object is made, often by reference to such a drawing.

Two such objects are semiconductor chips and the masks used during chip manufacture. Design right extends to chips and masks, and is dealt with separately in Section 3.6.

3.2 Design right

3.2.1 Introduction to design right

The concept of design right was introduced into the UK in the Copyright, Designs and Patents Act 1988 (the 88 Act) which came into force on 1 August 1989. The intention, made clear in parliamentary debates, was that the multitude of items made by industry should be legally protected in recognition of the substantial design effort expended on them.

The legal concept was totally new, not only in the UK but also worldwide. The definitions are also new, some of them having no equivalent in other intellectual property (IP) law. At the time of writing this book, there have been few decisions in the courts, so interpretation of the wording is as yet uncertain. Even when the wording duplicates that used in another IP law, there is no guarantee that the same meaning will apply to design right, but looking at the wording sometimes helps with interpretation.

3.2.2 What is protected by design right?

Design right applies to articles, that is, to tangible three-dimensional objects. There is no limitation on size; the right applies from micro-miniaturised components up to the largest size of transformer. By analogy with registered designs law, 'an article' probably includes a ship, an aeroplane and a prefabricated building. The fact that the article is subsequently attached to the floor or the ground is no bar, so long as it could be transported elsewhere, whether in knock-down form or as a whole.

The overlap with buildings which are protected by copyright is uncertain in the engineering context – is a microwave tower an article or a building? The 88 Act is unclear, so we must wait for a relevant court decision before there can be fully authoritative comment. What is certain is that design right applies to the vast majority of objects manufactured in the UK.

The first requirement for protection is that the design must be original; this is defined as being 'not commonplace' in the design field in question. There have been several cases in which the words have been considered, and one case reached the Court of Appeal. The product was an agricultural slurry separator. It was held that not only was the machine as a whole 'not commonplace' but that several specific basic parts of it were also not commonplace, such as the inner hopper that had a flat bottom, steep sides, a specific capacity and an 110 mm outlet hole. It also had a scraper with a spring loaded hinge.

The test applied by the court was to compare objectively the similarities and differences between the design of this slurry separator and of other slurry separators available at the time of creation of the design. The closer the similarities, the more likely it is that the design is commonplace, and there may be only one way in which such an article can be designed. On the other hand, the objective of this law is to protect the design of functional articles, and if it is original in the sense that it is not a slavish copy, then design right will apply. Even though a hopper for a slurry separator is a commonplace article, a particular design of hopper can be 'not commonplace'.

Design right applies to the 'shape or configuration' of the article; until 2001 this phrase was used in registered designs law (see Section 3.3) and, from legal decisions in that context, it is difficult to find any difference in meaning between the two words. Both internal and external shapes are protected by design right, so an article with its exterior identical to a previous design but differing internally is protected. Parliamentary debates made it clear that the internal layers of a semiconductor chip are to be included, so the wording of design right law should apply even to an internal feature which cannot be seen.

If the article is in the form of a kit, for example, an article for the DIY market, design right still applies, both to the whole article when assembled and to each part of it. The kit needs to be only 'substantially complete', so the omission of standard items, such as fixing screws, would not invalidate the legal protection.

3.2.3 Exceptions to design right

3.2.3.1 Surface decoration

This is self-explanatory. Articles on which surface decoration can be important, such as a coffee mug bearing a printed pattern or having a textured surface, can be protected in so far as decoration is concerned by registered designs law, but not by design right.

Not every surface feature is excluded. The connector pattern on a printed circuit board (PCB) is functional, not decorative. A PCB would be protected by design right provided the pattern of the connections was not commonplace, which in context probably means that either the circuit or its layout is new, or a substantial part of the circuit is new or rearranged.

3.2.3.2 A method or principle of construction

In a dispute relating to leather cases for mobile phones, the judge commented that stitching the seams was a method of construction and therefore excluded from design right.

3.2.3.3 The 'must fit' exception

Legal protection does not extend to the part or parts of an article which connect it to another article. For example, if a company designs a new type of fuse holder, the parts of the holder that actually make contact with the fuse can be copied freely by other manufacturers, although the rest of the fuse holder could not be copied without permission.

The intention of this legislation is that spare parts for a product should be available from other manufacturers. The wording of the exception was devised with car exhaust systems in mind, based on a legal action in 1986 by British Leyland against Armstrong Patents Company Limited relating to replacement exhaust systems for British Leyland cars. Armstrong sold the systems at a lower price than British Leyland. British Leyland sued for infringement under the law then applying (which was copyright law) and the case went all the way to the House of Lords. British Leyland won the argument that copyright had been infringed, but the House of Lords held that everybody had the right to have their car repaired at a reasonable price, and copyright could not be used to override this right. When the 88 Act was drafted, this position was kept in mind. Under the current law, anyone can make exhaust systems for a car and may copy the parts which fit to the car, using the 'must fit' exception. The rest of the exhaust system must be of a different shape to the original which is protected by design right.

One case under the present law considered the phrase as applied to contact lenses. The judge decided that the rear surface of a contact lens had to fit the eyeball, the front surface had to fit under the eyelid, and the lens diameter was chosen to fit onto the eyeball, so all of these features were excluded from design right protection.

3.2.3.4 The 'must match' exception

If the article in question forms an integral part of another article, then it is not protected by design right at all. Again, cars provide a good example. A replacement car wing or car door must match both the previous wing or door and the one on the other side, because the car was designed as a whole. Under this exception, it appears that anyone can make and sell replacement parts of this type.

3.2.4 When does design right apply?

The protection applies automatically; there is no need for formal action and nowhere that the right can be registered. Legal protection is available from the moment the design is recorded, whether it is created on paper or on screen. It also applies when the design is recorded as a written description or as a collection of numerical data stored on a computer. Making a mock-up or prototype also initiates protection. A photographic record can also form the basis for protection, but presumably there must have been an article to photograph – possibly a digital photograph could be altered to create a new design. All the types of records listed above are known as 'design documents'.

The period for which protection lasts depends on whether the design is put to practical use. If the article is made and sold or hired out, or even advertised, design right lasts for 10 years from the end of the year in which the sale, hire or advertisement occurred; that is, if the article was advertised for sale in July 2005, design right lasts for 10 years from 31 December 2005 to 31 December 2015. If the first sale occurs close to 1 January, almost 11 years of protection will result, but other factors may affect the first marketing date more strongly than the aim of a maximum term of legal protection. The owner of the design does not have to make the sale or hire directly, but must authorise it.

If the design never gets off the drawing board or is never developed beyond prototype stage, legal protection lasts for 15 years from the end of the year in which the drawing or prototype was made.

3.2.5 Who owns design right?

This question is really in two parts:

- Who is the designer?
- Who is the owner of rights in the design?

The first part is simple to answer, the second is less straightforward.

3.2.5.1 The designer

The engineer who actually creates the design is the designer. For complex articles, there can be more than one designer, and a joint design would result.

In addition to a human designer, the 88 Act envisages a design being created by a computer without human intervention. This does not mean the use of a computer-aided design program, when the engineer has a supervisory function and the process is interactive, but means a design made without any human control. This could, for example, be a complex PCB layout where component lists are entered into the computer which then follows preprogrammed rules to devise the best physical arrangement for the circuit and the connections. In such a case the design right is owned by 'the person by whom the arrangements necessary for the creation of the design are undertaken'. Since in legal terminology 'a person' can be a company, probably the company owning the computer on which the program is run would own the right.

3.2.5.2 The owner

If the designer is an employed engineer who produces the design as part of his or her job, then the engineer's employer owns the design right (see Chapter 7 for more detail).

If the design is created under a commission, that is, one company pays another company or an individual to make the design, then the company paying for the work automatically owns the design right. The commission must be either for money or for money's worth. The latter phrase means that there could be a complex financial agreement between the two parties, of which creation of the design forms a small part, with money actually changing hands in the reverse direction to the commission instruction.

The two sets of circumstances summarised above apply to the creation of design within the UK by a British subject. If the design is created in another country, then UK design right exists if the individual designer, their employer, or the commissioner of the design, either lives in or has a place of business in certain other countries, currently any EU country. This provision may be extended to any country that gives protection to designs on a reciprocal basis. So far, it applies to the Channel Isles, the Isle of Man and New Zealand. Some countries operate a registration system for functional designs, as well as aesthetic designs of the type which can be registered in the UK, and this might be the basis for arguing reciprocal protection. Other countries also use laws relating to unfair competition to protect designs.

If neither the designer, nor the employer, nor the commissioner live or have a business in the UK or other EU country, then protection can be generated by first marketing the designed article in an EU country, provided the owner authorises such an offer for sale or hire. If an article designed in the USA or Japan and never marketed in the EU is copied in the UK, it seems the designer would have no basis for legal action under design right law.

3.2.5.3 Transfer of ownership

After its creation, design right can be transferred into new ownership, but the transfer must be recorded in writing in an assignment document which must be signed by the owner making the transfer.

Ownership can also be transferred in advance, again on condition that there is a signed document. For example, an engineering designer could have been commissioned to produce one design, and might agree that all similar designs of the same article generated in future would also belong to the commissioner, even though the commission would be regarded as complete when the first design was ready. As with employee-created designs, the assigned-in-advance right would never belong to the designer, but would be the property of the assignee from the moment of creation.

3.2.6 Example of design right

The law set out above is well illustrated by a familiar object, the traffic warning light sold by Dorman Traffic Products Limited under the name 'TrafiLITE' (registered trade mark). Figure 3.1 shows one version of such a light, which has a polycarbonate

Figure 3.1 'TrafiLITE' warning light

lens on an impact resistant polypropylene body. Figure 3.2 shows the lamp carrier from inside the body adjacent to the body. Clearly, even an apparently simple device requires considerable design effort.

The illustrated version was entirely designed after 1 August 1989 so design right applies, although there were earlier versions and also a version in a metal case. The assembled lamp is an article, and each component part of it is also an article, and as such each is protected by design right.

The polycarbonate lens is computer designed, so the design document for it is in the form of a computer record. The body and the lamp holder were designed on paper, and the paper drawing is the design document. All of the designs were produced by

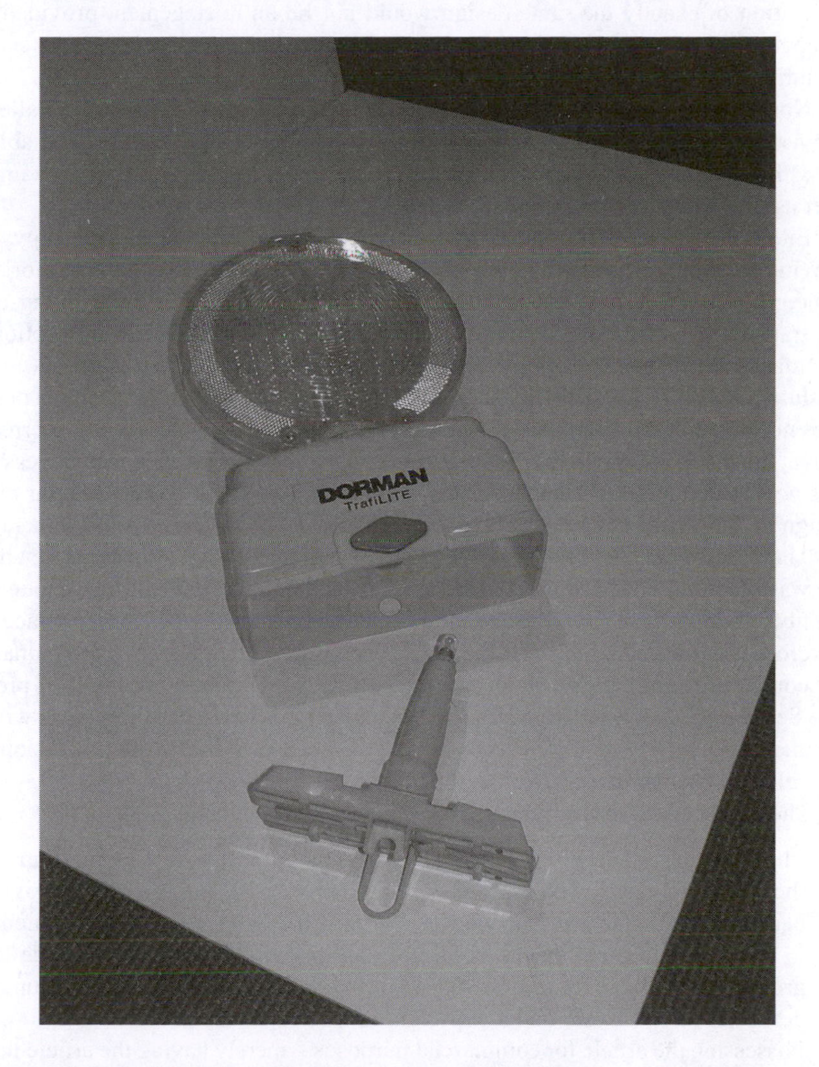

Figure 3.2 'TrafiLITE' warning light – component parts

employees of Dorman Traffic Products Limited so the company owns the design rights. The computer program to design the lens was also written in-house by an employee, so copyright for the program also belongs to the company.

3.2.7 Infringement of design right

Owning design right in an article allows the owner to stop others from competing in several different ways.

The most important right is that the owner can prevent anyone else from copying the design and making exactly the same article or one which differs only slightly. This is known as primary infringement. The copying aspect is important; independent generation of exactly the same design would not be an infringement provided the competitor could prove this, for example, with dated drawings showing independent evolution of the design.

Normally, the onus will be on the owner of the right to show that the alleged copyist had access to the design in question. For example, the owner may be able to show that an ex-customer has acquired the articles from another source, for example, after asking another manufacturer to make copies.

Either the article itself or the design document can be used to make the copy. The copying need not have been actually carried out by the other person, and authorising someone else to copy is also legally actionable: subcontracting the copying does not transfer the legal burden to the subcontractor if there is a clear instruction to manufacture the copy.

Just because the allegedly infringing article is almost exactly identical to a design does not prove that there had been copying. In the case of contact lenses referred to above, the alleged infringer had fed figures, such as lens diameter, lens thickness and lens power, into their own spreadsheet. The resulting lenses were very similar to the design in question, but there had been no copying and therefore no infringement.

There is a legal defence to actual copying: the alleged copyist can argue that he or she was not aware that design right existed, that is, that infringement was 'innocent'. Possibly, the argument would be that the design was presumed to be commonplace and therefore unprotected. If this argument succeeds, the court will not award damages, but could still award an injunction to stop further copying or an account of profits (see Section 9.2.2). General ignorance of the law cannot be argued; engineers will be assumed to know the basics of design right law especially if they work professionally in a relevant business area.

The owner can also stop a range of secondary infringements:

1. Importing the article into the UK for commercial purposes – this applies if making the article in the UK would have been an infringement, and is a useful provision because a manufacturer outside the UK is difficult to sue. Even an intention to import is an infringement, so placing an order is sufficient to provide legal grounds for action. Importing articles for private and domestic use is presumably permitted as is the case under copyright law.
2. Possessing the article for commercial purposes – merely having the article under your control with an intention to trade with it is an infringement.

3. Selling it, hiring it out or offering it for sale or hire as part of a business – it is easy to identify who is trading in an article, locating its manufacturer is much more difficult.

A condition is attached to these rights to sue secondary infringers; the infringer must know that the article is an infringement of design right. Engineers, manufacturers and businessmen in general are presumed to know the parts of the law applying to their business.

If the design right owner succeeds in proving infringement, the court can award an injunction to stop further manufacture or import, order payment of damages (e.g. for loss of sales by the design right owner), and/or the destruction of the offending articles produced during the previous 6 years or handing them over to the design right owner (see Section 9.2.2 for more detail). If the design right owner was prevented from taking action within the 6 years by some act of fraud or concealment of facts, the time limit can be extended. The 6 years is set by the Statute of Limitations and is a general time limit on when a legal action can be started.

Additional rights apply to articles such as the moulds for producing an infringing item. If the person owning or controlling the mould knows it is intended for a use which infringes design right, the court can order the mould to be delivered up to the owner of the design right to prevent further infringing use. Returning to the TrafiLITE, the traffic warning light example, if Dorman Traffic Products Limited learned that a mould had been designed to manufacture copies of their lenses, an order to deliver up the mould (a very expensive item) would be an effective way of stopping the unauthorised copying. But the mould would have to be capable of producing lenses that are identical or very similar. If the lenses are of a different design but fit onto the Dorman body, their manufacture or sale cannot be prevented because the 'must fit' exception applies.

If the design right owner has granted an exclusive licence to a third party, that is, permission for a third party to make the article when the owner agrees not to manufacture in competition, the exclusive licensee has the same right to bring an infringement action as the owner.

Legal protection also extends to copying design documents to allow manufacture of the article. The relevant parts of the 88 Act have been the subject of much legal speculation, but there is little doubt that such copying would be an infringement under copyright law with its much longer period of application (designer's life plus 70 years) than design right law. Making a new drawing by starting from the article itself may infringe such copyright also, but a court decision on the point is needed for certainty.

For total safety, the message is 'Don't copy anything unless you are sure legal protection has expired'. Since design right applies to precise copies and substantially similar articles, an object can be redesigned so that its shape is distinctly different when made, and thus would no longer be an infringement at any time.

3.2.8 Exceptions to infringement of design right

There are two major exceptions; one relates to the last 5 years of existence of design right, the other concerns use by the Crown.

3.2.8.1 Licences of right

Although design right lasts for 10 years from first marketing of the article or for 15 years after preparing a non-marketed design, the owner does not have the full right set out above for the whole of those periods. For the last 5 years of the 10 or 15 year period, the owner cannot stop competitors from manufacturing articles to the design, because a licence of right is available under the 88 Act. This means that the owner is regarded as having given permission for manufacture but, as is normal in licence agreements, the licensee is obliged to pay for the permission in the form of a royalty. Such licences differ from normal licences because the licensee is not allowed to advertise the trade connection with a design right owner, unless the owner gives permission to do so.

If the owner and the licensee cannot agree on the licence terms, then the Comptroller of Patents at the patent office will set the terms. These can be financial and non-financial but it is expected that royalty rates will be the subject of most disputes referred in this way. The Comptroller will not consider an application to settle terms unless the licence of right period has already begun or will begin within 12 months of the application. The Comptroller also has the power to grant a licence if the owner of the design right cannot be identified.

A practical problem for those wishing to use rights belonging to other companies is that since design right is not registered, it may be difficult to know when the design was first offered for sale, and therefore to work out when the licence of right is available.

3.2.8.2 Use by the Crown

A government department may copy a design and this is not regarded as infringement. The government may also authorise other companies to manufacture the design for use in the defence of the UK or for foreign defence purposes or for use in the health service – this is held to be the pharmaceutical, general medical and general dental services. The right extends to selling surplus stocks.

If the Crown authorises use of a design in this way, the owner of the right must be notified and informed of the extent of use. The government department in question pays compensation to the owner for the loss resulting from manufacture of the articles by another company, but the compensation is limited to the level of supply which the owner could have fulfilled from the company's manufacturing facilities at the time. For example, if a small engineering company of limited manufacturing capacity designs an article which is used in enormous numbers by a government department, the small company will not be compensated for the whole use.

3.2.9 Threats

If the owner of a design right believes the design is being misused, the immediate reaction might be to write to the supposed infringer, telling them to stop and threatening to sue them if they do not comply: this can be dangerous. Some persons threatened in this way have the right to sue the design right owner if the threat turns out to be unjustified, and can obtain a formal declaration to that effect, plus an injunction against further threats and damages for any loss suffered.

The right's owner is permitted to notify the supposed infringer of the existence of the design right, without saying what the next step might be. Unfortunately, in communications with small companies, the response is often puzzlement. Without knowledge of the legally allowed format, the recipient of the letter often thinks 'So what?'. But such a notice must be sent before any subsequent letter referring to the possibility of legal action.

Fortunately for the right's owner, only a limited number of people can bring a threats action. It cannot be initiated by anyone who is making or importing articles which are believed to be infringements, so explicit letters can be written to such competitors who are usually the main problem for design right holders. Only those who are secondary infringers, for example, those who possess infringing articles in the course of business, or who sell or hire them out, are considered to need protection from threatening letters. In this way, customers of manufacturers who copy a design are protected from threats which should be directed towards the manufacturer.

A threat does not have to be in writing; a verbal threat can be the basis for a threats action, but inevitably it is more difficult to prove. If a threats action is brought against a design right owner, a complete defence is to show that the articles were indeed infringements. The safest course of action when suspecting infringement or on receipt of a threatening letter is to seek advice from a patent attorney.

3.2.10 The position in other countries

Some countries protect functional designs by a registration system, while others apply copyright law or unfair competition law. In the EU there is a Community Design Right, see Section 3.5.

3.3 Registered designs

3.3.1 Introduction to registered designs

A disadvantage of the formality-free design right is that the owner must prove copying. The advantage of registered design is that showing the designs are the same or very similar is enough. Registration can last for 25 years, provided conditions are met and fees are paid.

The relevant law is the Registered Designs Act 1949 which was amended by the Copyright, Designs and Patents Act 1988 and again in 2001 to meet EU requirements. The same law applies or will soon apply in all 25 EU countries.

3.3.2 What does a registered design protect?

The law provides a comprehensive definition:

'design' means the appearance of the whole or a part of a product resulting from the features of, in particular, the lines, contour, colours, shape, texture and/or materials of the product itself and/or its ornamentation.

So a design can be two or three dimensional and it can be the whole or a part of a product, which is also defined. 'Product' means any industrial or handicraft item, including packaging, get-up, graphic symbols and typographic typefaces but excluding computer programs.

This means that a product can be something that is mass-produced or it can be a one-off item such as a sculpture. Computer icons are included. Protection also extends to parts which are assembled into a complex product, that is, one which can be assembled and then disassembled, such as temporary seating.

Internal features can be registered, such as the inside of a suitcase, but products such as vehicle spare parts which are not seen in normal use are excluded. The Designs Registry gives the example of a replaceable ink cartridge in a pen.

3.3.3 Novelty and individual character

For an application to be valid, the design must be new and have an individual character.

The novely requirement is absolute; if the design is available anywhere in the world before an application is filed in the Designs Registry of the UK Patent Office, it is not new. The prior design does not even have to be applied to the same product. So if a pattern is known for use on furnishing fabrics, it cannot be registered for wallpaper. The earlier design does not have to be completely identical, it can differ in 'immaterial details' and still knock out the later application.

There are two exceptions. The first is that the designer is allowed a 'grace period' of 1 year in which to test out the market, or try to get funding, and can disclose for these purposes, then apply to register within 12 months of first disclosure. The second is that if the first publication could not reasonably have been known to the applicant in the normal course of business, then it does not count. Possibly this means that a publication or use in an obscure country, perhaps a decorative article on display in a small museum, is excluded. This is unlikely to apply to engineering articles.

The test for individual character is that the design must produce on an informed user an overall impression which differs from the overall impression produced by the earlier design. 'Informed user' is not defined but commentaries make it clear that the user is not a skilled designer. The Designs Registry gives the example that for a coffee pot, an informed user would be a waitress. Will the new design produce a *deja vu* reaction when the waitress sees it?

In judging individual character, the designer's degree of freedom must be considered. This recognises that for some products there are many constraints, while for others a designer has complete freedom.

3.3.4 What is excluded from a design registration?

If a feature of a product is dictated by its function, then it cannot be protected by a design registration. This is very similar to the previous law where the point was considered in court in a case relating to electric terminals. AMP Incorporated had

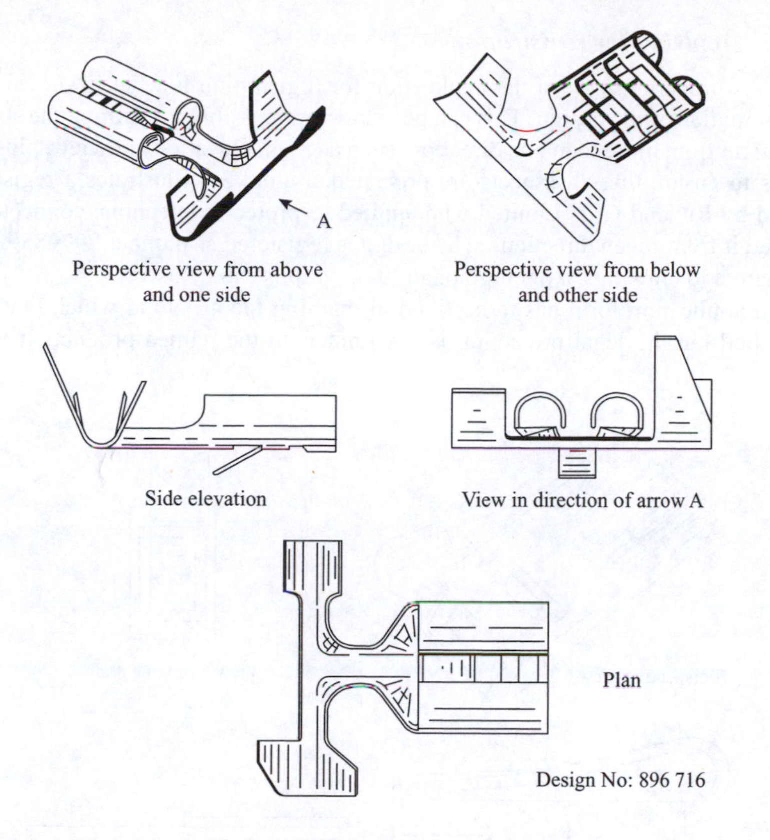

Figure 3.3 Electric terminal of AMP Incorporated

a registered design (Figure 3.3) for a terminal which had a receptacle, a narrow connecting neck, and a connecting channel at right angles to the receptacle. The channel had two sets of ears which, when used, were crimped round a wire to give a flag or battle-axe shaped connection. A competitor, Utilux Proprietary Limited, started to sell very similar terminals and AMP sued them for infringement of the registered design. The arguments related largely to the shape of the ears, and the case went through the appeal process up to the House of Lords. The decision was that the shape of the ears was dictated solely by their function so the design registration was invalid.

Another exclusion from registration is a feature which allows a product to fit to another product, which is similar to the 'must fit' exclusion from design right, Section 3.2.3.3. However, this does not apply to design features that allow modular systems to be fitted together – such features can be registered.

A design application already in the system but not yet published, and therefore not destroying novelty, can also invalidate a later application if the two designs are very close.

3.3.5 *Applying for registration*

The most important part of the application for registration is what is known as the representation of the design. This can be a drawing, possibly illustrating the shape of the article from more than one direction, such as plan, elevation and view, plus other angles to ensure that all aspects are protected. Figure 3.4 illustrates a registration owned by Rutland Gilts Limited who applied to protect a telephone connector and showed it from seven directions. The design is registered as number 2079859. As an alternative to drawings, a photograph or photographs can be used.

An application form has to be filled in, naming the article to which the design is applied but the legal protection is not limited to the named product. It is used

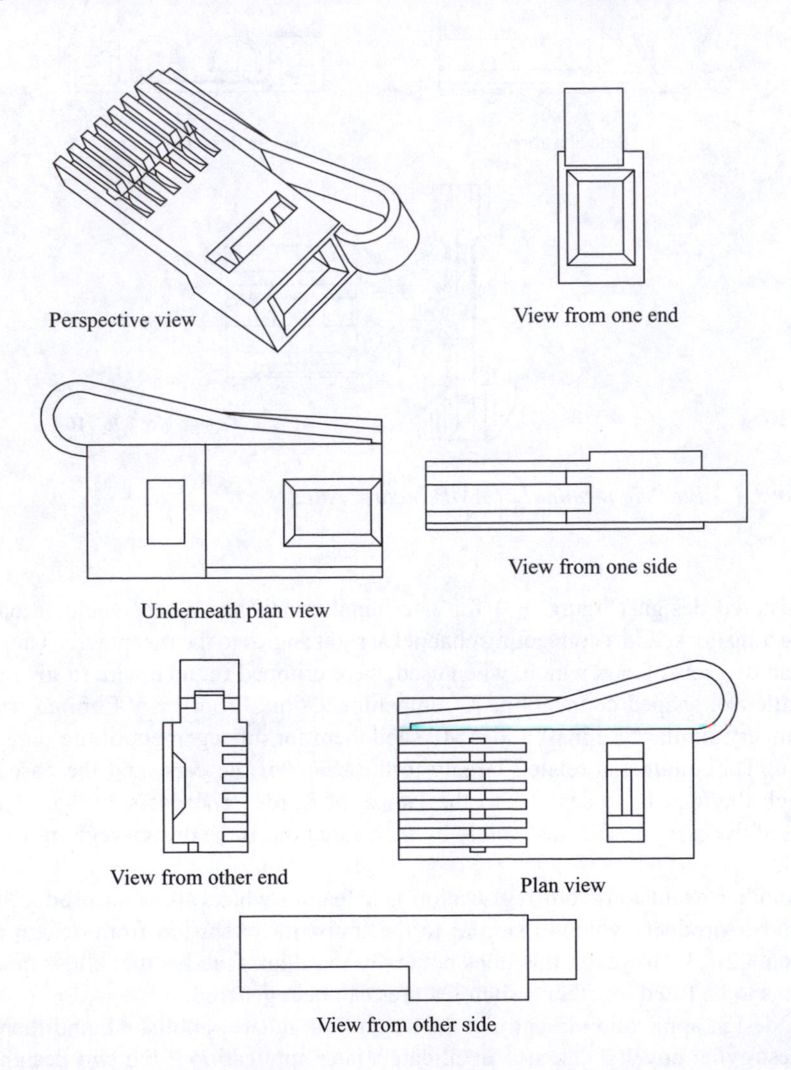

Figure 3.4 Registered design for a telephone connector

merely for classification purposes using an international system, the Locarno system, which puts all products into classes such as 'Class 10 Clocks and watches and other measuring instruments, checking and signalling instruments' or 'Class 13 Equipment for production, distribution or transformation of electricity'. The entire classification can be accessed at www.patent.gov.uk by selecting Designs followed by Reference material and then Locarno Classification.

Naturally, a fee must be paid (which at the time of writing this book is £60) and various other items of information given. The application can be dealt with by a patent attorney who will be familiar with the formalities required.

The representations, forms and fee are all filed at the Designs Registry, which is part of the patent office, and the application is date-stamped on the day of receipt.

An examiner then checks that the application is formally in order, that is, it applies to something which on the face of it is a design of a product and is not offensive and does not include national flags or portraits of the royal family, and that it is not limited by technical function. But the examiner does not check whether the design is new. It is rejected only if it is self-evidently not novel, or if the examiner just happens to know of a very similar design already registered.

If the applicant wants to check novelty, the registry will search existing registrations for a fee of £25. This can also be done before the application is filed. Usually registration takes about 3 months. A certificate is then issued and the registration is published. The owner now has the right to start proceedings for infringement.

The initial registration period lasts for 5 years from the application date. If renewal fees (in the low hundreds of pounds) are paid, the registration can be kept in force for four subsequent periods of 5 years, totalling 25 years.

Successful registration may not be the end of the story. 'Any person interested', such as someone with a similar design, can apply to have the registration cancelled or invalidated on the grounds that the design was not new or, if it was registered after 9 December 2001, because it lacked individual character. Such challenges are rare but the possibility exists.

3.3.6 Who owns the right?

The question falls into two parts:

- Who is the designer?
- Who owns the right?

In most cases answering the first question is simple; it is the draughtsperson who made the drawing. The statute law assumes that only one person creates the design, there is no provision for joint designers.

There is another possibility. The UK law deals with the situation when a shape or pattern is created by a computer. This does not refer to use of a computer-aided design program with a human being in control of the process, but refers to creation without human intervention. In these circumstances the company owning the computer in which the program is run would almost certainly be regarded as the designer.

The second part of the question, ownership of the right, has several possible answers.

1. The designer is an employee – if an employed designer creates a design as part of his or her job, then the employer is treated as the proprietor of the rights. For more detailed consideration of the point see Chapter 7.
2. Creation under commission – if a self-employed designer is commissioned to design a product, or if one company commissions another to do so and an employee creates a design, then the company placing the commission automatically owns the right to register the design. The commission must be for money or money's worth.
3. Assignment – there can be an agreement, preferably in writing, transferring ownership of the right to apply for a design registration. This can be effective either before or after the application is filed for the registration of a design.

3.3.7 Infringement of a design registration

The design owner has the exclusive right to use that design and any design which does not produce on an informed user a different overall impression, that is, the test is the same as for individual character.

The test is purely visual. The allegedly infringing article is compared with the representation in the registration, and can also be compared with the actual article the registration protects.

It does not matter that the product is different to that stated on the application form, any product using the design infringes the registration. So a registration of a design for a car body will be infringed by a toy car, a china ornament, a print on a teatowel, or a chocolate bar if any of them use the same design.

It is an infringement to make, offer, put on the market, import or export, or to stock it for such a purpose.

A great benefit of a design registration is that it is a true monopoly right. There is no need for the owner to show that there has been any copying, it is sufficient that the same or a substantially similar design is being applied to a product. Even if the infringer proves that the design was created independently, there is still infringement.

Design registrations are publicly available and requests can be made to the registry to search for similar designs before a new product is marketed, or before the major expense of tooling is incurred, therefore the system is not unfair.

There is a defence: an alleged infringer can argue that the infringement was innocent, that he or she did not know that the registration existed. However, businessmen and women are expected to have a general knowledge of the law relating to their area and genuinely innocent infringement is difficult to prove. It is up to the person accused of infringement to prove their innocence and obviously this will not be the case if the goods have been copied. If the alleged infringer can prove innocence there will be no award of damages by the court, but there could still be an injunction to stop further use of the design.

Another exception is that if the goods have been put on the market in any country in the European Economic Area (the EU plus Iceland, Norway and Liechtenstein) by or with the consent of the owner, then there is no further control by the owner who cannot stop those goods being moved to another country and sold there at a different price.

An infringement action cannot be started until after a design has been registered, a pending application is not sufficient. Since registration usually takes only about 3 months, the delay is rarely serious in practice.

Even if the application has been successful and the design is registered, the owner needs to approach a suspected infringer with care. Writing a letter threatening to sue can have severe consequences. If there is infringement there is no problem but if it turns out that there was no infringement of the registration and that the person was not making or importing the articles but was just trading in them, the person threatened can sue the owner for making an unjustified threat and can be awarded, among other things, payment of damages.

What the owner can do is draw the alleged infringer's attention to the existence of the design – if a letter does only this, it does not constitute a threat. A threat does not have to be in writing, it can be verbal. A threat need not be made in a letter to a person or company, a general statement or advertisement can also form the basis of a Threats Action by anyone who feels they have been specifically identified. The safest course is to seek advice from a patent attourney before taking any action at all.

3.3.8 Designs and use by the Crown

Registered designs relating to certain classes of articles may be the subject of a secrecy order, and any such design can be used by a government department for defence purposes. The regulations are similar to those for design right, as explained in Section 3.2.8.2.

3.3.9 Marking

There is no obligation to mark an article to show that it is protected by a registered design, or to mention this in descriptive literature, but such a marking does have the advantage of acting as a warning. Infringers cannot then argue that they infringed innocently, but it is essential for the registration number to be used. Marking it with just 'registered' or 'registered design' without the number is regarded as insufficient legal warning that a registration exists. The interested public must be given the full information so that the design can be checked and therefore its date of expiry or non-renewal determined.

If the article itself or product brochures etc. are marked with the number of the registration, a system must be put in place to make sure that the marking is discontinued when the registration either runs out after 25 years, or a decision is taken not to pay a renewal fee. If the marking continues, this will be a false representation that the design is registered, a criminal offence which can result in a fine.

3.3.10 Old law

The law in force before 2002 was based on the Registered Designs Act 1949 as amended by the Copyright, Designs and Patents Act 1988 which came into force on 1 August 1989. A design had to 'appeal to the eye' as well as being new and there was no test for individual character. A registration applied only to the article named on the application form. Designs under the old regime can, if all renewal fees are paid, run until 2014.

3.3.11 Registrations overseas

As with other intellectual property rights, design registrations are granted in other countries under national laws. Many other countries have a registration system and if protection abroad is required an international convention assists the process. If a design application is filed in the UK and an application is filed in any country which is a member of the convention within 6 months, then the effective date in that country is the UK date. The convention application will take precedence over any actual local application filed after the UK application date.

Most industrial countries which have a design registration system are members of the convention. Other countries do not have such a system and protection for designs varies much more widely than is the case with copyright or patents.

For all EU Member States the rules are, or soon will be, identical to those in the UK.

Some countries with registration systems limit the protection to articles which appeal to the eye, but a few allow totally functional articles to be registered. Countries which have no registration system at all rely on other law, such as unfair competition law.

The USA also has a registration system; the protection is known as a design patent ('conventional' patents are referred to as 'utility' patents to distinguish them). Design patents are only granted if the article is the product of aesthetic skill and artistic conception, they are examined to check that the design is novel, and they last for 14 years from the date of grant. A US design patent requires a description and a claim, but these can be minimal, as shown in the example of Figure 3.5 which shows US design patent number 310, 120 for a medical monitor.

Design registrations are also available in Japan. The design must be novel and have aesthetic features. A registration lasts for a maximum of 15 years.

3.4 TRIPS and industrial designs

The TRIPS Agreement (TRIPS – Trade-Related aspects of Intellectual Property Rights) places on countries, which have signed it, the obligation to provide legal protection for industrial designs, that is, for objects intended for mass manufacture. The design must be original, meaning that it is significantly different from existing designs for the object, or for a pattern to be applied to it. The legal protection must not extend to how the object functions.

The owner of a protected design must have the legal right to prevent third parties from making, selling or importing the object when made to that design, without the owner's permission.

The minimum provision is for the protection to last for at least 10 years, which may be divided into two periods of 5 years.

United States Patent [19]
Wickham et al.

[11] **Patent Number: Des. 310,120**
[45] **Date of Patent: ** Aug. 21, 1990**

[54] **MEDICAL MONITOR**

[75] Inventors: John F. Wickham, Northamptonshire; Trevor A. Nunn, Witney; Derek S. Jay, Wiltshire, all of United Kingdom

[73] Assignee: Oxford Medical Limited, Oxford, England

[**] Term: 14 Years

[21] Appl. No.: 135,283

[22] Filed: Dec. 21, 1987

[52] U.S. CL D24/17; D24/21

[58] Field of Search D24/17, 21, 8; 128/668, 128/670, 671, 680, 681, 689, 696–716, 736, 900, 906; D10/46, 76

[56] Reference Cited

U.S. PATENT DOCUMENTS

D. 264,128 4/1982 Barnes et al. D24/17
D. 296,240 6/1988 Albright et al. D24/17
D. 299,861 2/1989 Hunsdale et al. D24/17
4,121,574 10/1978 Lester 128/666
4,404,974 9/1983 Titus 128/670
4,724,844 2/1988 Rafelson 128/670

Primary Examiner—A. Hugo Word
Assistant Examiner—Stella M. Reid
Attorney, Agent, or Firm—Martin Novack

[57] **CLAIM**

The ornamental design for a medical monitor, as shown.

DESCRIPTION

FIG. 1 is a top, left side perspective view of a medical monitor, showing our new design;
FIG. 2 is a rear elevational view thereof;
FIG. 3 is a right side elevational view thereof;
FIG. 4 is a bottom plan view thereof; and
FIG. 5 is a top, left side perspective view thereof; in an alternate position, the bottom being substantially plain and unornamented.

Fig. 1

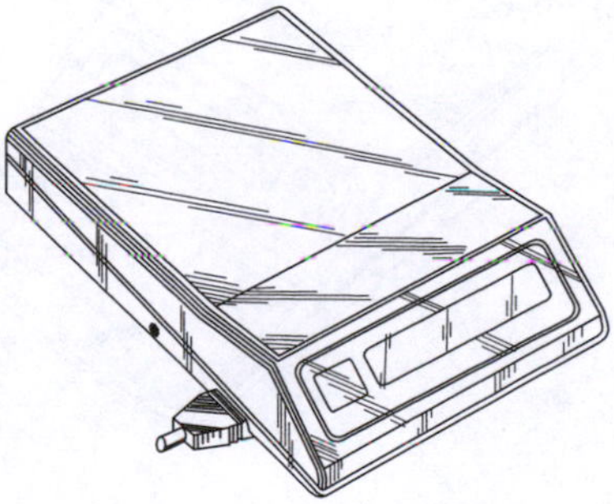

Fig. 2

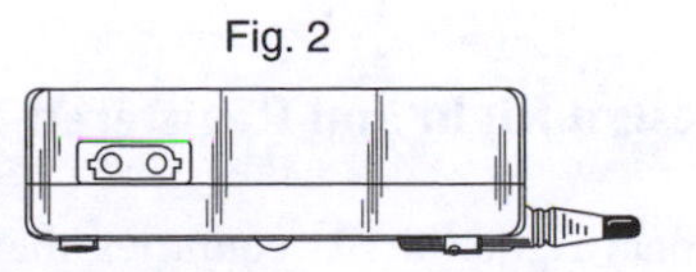

Figure 3.5 A US design patent

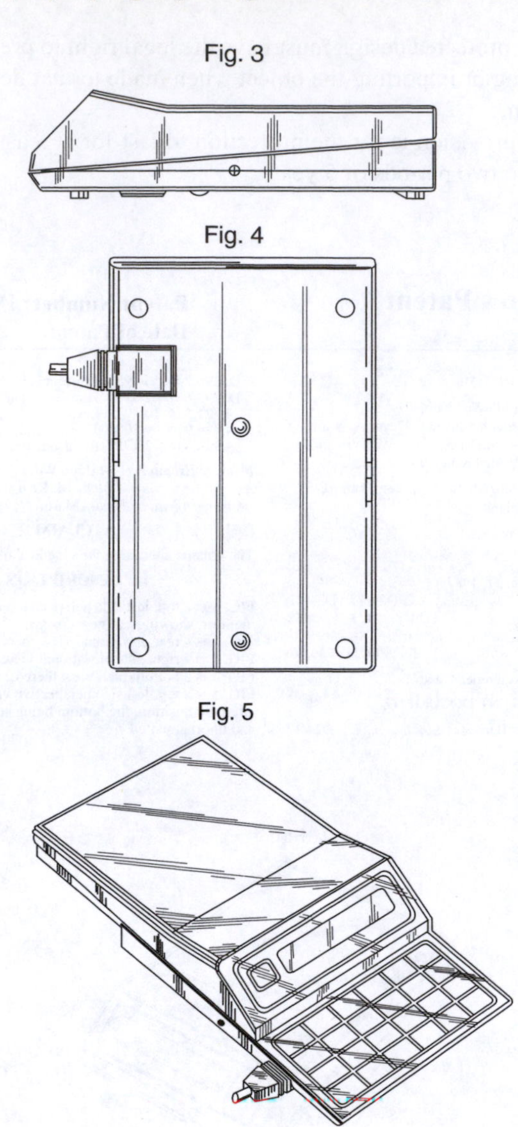

Figure 3.5 (Continued)

3.5 Community Design Right and Registered Designs

In addition to the national rights in EU countries there are two rights which are unitary, that is, they each apply as a whole to all EU countries and cannot be divided nationally in any way. One is an unregistered design right which came into force in March 2002 and the second is for registered designs, starting on 1 April 2003. Both are administered by the Office of Harmonisation of the Internal Market based in Alicante, Spain.

The advantage of a Community right is that it has the same effect in all EU countries, and that for the registration system there need be only one application. The disadvantage is that if the protection fails for some reason, it fails in all countries.

The unregistered right lasts for 3 years from first publication. The registration is for 5 years from the date of application or the priority date if it originates outside the EU, renewable for four further periods of 5 years.

For both rights the definitions of 'design' and 'product' are identical to those in UK registered designs law: there are the same requirements of novelty and individual character, the same exclusions of, for example, technical function and the same tests for infringing acts.

For the registration, an application can be filed in Alicante or in a national patent office which transfers it to Alicante. There is no examination for novelty, but there is the possibility of filing multiple designs provided they are all intended for use in the same class of product in the Locarno system (see Section 3.3.5). The first design attracts the normal fee of €350 while the second and all further designs attract lower fees. The multiple application can be divided later or can proceed as a whole. A single or multiple application can also, for a fee, be deferred for up to 30 months and is not published at that time. The aim is that there is temporary protection for large numbers of short-lived designs allowing final commitment and investment to be delayed. Designs not proceeded with can be abandoned.

Each Member State must nominate a Community Design Court which will deal with any disputes and challenges. In the UK this is currently the same courts that deal with patents, see Section 9.2.3.

Community Registered Designs can be searched at www.oami.eu.int/en/design and there is an application form on this site also. The register is published in all official languages of the EU.

Having a Community right does not override a national right; both can exist simultaneously. The only limitation is that legal actions on the same design against the same third party cannot run on a national right and a Community right at the same time.

It is still too early to judge the success of Community rights. For registration the fee of €350 is not high and compares very favourably with the total fees for applications in all 25 countries. Certainly, applications in national offices have fallen markedly since the Community registration system came into force.

3.6 Topography rights

3.6.1 Introduction to topography rights

Semiconductor chips are now ubiquitous. They are incorporated in a huge range of consumer products, from a digital watch to a laptop computer, and it is largely the chip that allows efficient performance and low prices. In 2003, chip sales totalled £100 billion worldwide with most production being in silicon factories located in the Far East.

The circuits within a chip can either perform the analog functions of amplification and modulation, or digital switching techniques can be used. A microprocessor may contain millions of transistors acting as gates or stores. A semiconductor chip can be a standard product, such as a microprocessor, ROM or RAM, or it can be custom designed, for example, for use in telecommunications. Between these two extremes there is a range of semi-custom designs. The basic layout of a chip is fixed, but a user can select the connections within it from a library of possible arrangements made available by the chip manufacturer.

The user selects features from a vendor library, locates them on an electronic drawing board and draws in the connections. This is the equivalent of a circuit diagram; it is not the layout of the chip itself, but it can be used to test the circuit logic, with a specifically written test program. When the circuit logic has been tested, the design in electronic form is sent to a chip manufacturer, who prepares the chip layout. This is known as an Application Specific Integrated Circuit (ASIC). Such a chip is cheaper than a custom design because manufacture is partly standardised, but the available space may not be used as efficiently as with custom chips because of the initial constraints.

Development of semiconductor chips is expensive. To design a new chip having 25 million transistors costs several million pounds. In contrast, copying chips is relatively easy. To do this, the casing is removed and the metal connection layer is photographed. This metal layer is then dissolved with acid and the top layer of the chip photographed. The camera is then focused through the translucent semiconductor material onto the underlying layers. The photographs can be used by a pirate company to copy the chip layer by layer at a relatively small cost, for example, £100 000 or so. The pirate company can therefore price the chip very much lower than the company trying to recoup its development expenditure.

In the early 1980s, copying of chips was widespread and the USA responded with a new type of law in 1984. The basis was that semiconductor products originating in other countries would be legally protected in the USA only if that country gave reciprocal protection to US designed chips. The USA set a deadline for compliance. In response, the EU issued a directive on the subject in 1987, providing further encouragement for Member Countries to introduce appropriate law. The first set of regulations in the UK was rushed through in 1987 to meet the USA deadline, and was amended later.

The legal rights discussed in this book are difficult to apply to chips. Many chips will be only slight variations of other chips, so patent protection is not possible because there is no invention. They are functional objects, so registered design and copyright are not applicable. Design right law is the best fit, but was drafted with larger, mechanical articles in mind. The applicable UK law is a compromise; design right applies, but is varied by additional regulations called the Semiconductor Products (Protection of Topography) Regulations 1989. The legal rights provided under UK law will be referred to as topography rights.

3.6.2 What is protected by topography rights?

The legal right is based on the definition of a semiconductor product and of 'topography'. A semiconductor product is defined as a product which has an electronic function and which has at least one layer of semiconductor material and at least one other layer. There must be a pattern on or in the semiconductor layer, and the pattern must relate to the electronic function. At its simplest, this wording includes a single layer of semiconducting material, doped in various areas to give a pattern and supported by an insulating layer; connections could be few in number and at the layer edges.

The definition excludes a conventional PCB because while a PCB has a pattern, that is, its electrical connection layer, it does not have a semiconducting layer. A PCB is protected by design right, not by topography right.

The second part of the definition, topography, refers to a pattern associated with the semiconductor product in some way. There are several possibilities:

1. The pattern is fixed in a layer of a semiconductor product, for example, it is the pattern of doping in that layer.
2. The pattern is fixed on a layer of a semiconductor product, for example, it is the metal connection layer on top of a semiconductor layer.
3. The pattern is intended for use as described in (1) or (2), but currently exists in another form, for example, on disk or on tape, even if it exists only as a set of coordinates from which a physical pattern can be generated.
4. The pattern is fixed in or on a layer of material in the course of manufacturing a semiconductor product, for example, the pattern is formed in/on a layer of material which is not itself a semiconducting layer and is not attached to such a layer – this could be the mask or stencil through which a photosensitive surface is exposed to light during manufacture of the chip, or a record of the pattern of movement of a laser beam directed onto such a surface during the manufacturing process.
5. The pattern is intended for use as in (4), for example, the pattern for a mask has been generated, on disk or tape, but the mask itself has not been made.
6. The arrangement of the layers of a semiconductor product in relation to one another, for example, the whole semiconductor product is protected, as any one layer by itself might not meet other requirements, such as originality.

So the scope of topography right is quite broad. Any pattern, however it is recorded, forming part of a semiconductor chip or to be used in manufacturing such a chip, and the whole chip itself, are all included within the definition.

3.6.3 What is not protected as a topography?

There are two exclusions: a topography which is not original and a topography which is commonplace.

The topography must be the result of the creator's own intellectual effort, and is protected only if it was specially designed and not copied from something else, although from the context it appears that the level of originality is not meant to be high. For example, consider topographies relating to semi-custom chips where there is a basic layout and a large number of possible circuit connections to be made. If only the basic layout without the connections is considered sufficiently original to be protected by topography right, then the extent of legal protection would be very limited. In contrast, if the required level of originality is low, then every set of connections made on the semi-custom chip which is different from a previous set is protected as an original topography. To give the intended effect of broad protection, the latter interpretation must be correct, but it has not yet been tested in the English courts.

Even if the chip was specially designed, if it is 'commonplace' it is not protected by topography right. That is, the pattern must not be one which is well known to designers of topographies or to semiconductor product manufacturers. It is not clear how many virtually identical products need to be available to render a topography 'commonplace'. A single example must be insufficient, but no minimum number is set. Again the courts will need to interpret the law before there can be any certainty.

The possibility exists of several commonplace patterns in different layers being combined to form a protectable, non-commonplace topography. This is covered by definition (6) above. Generally, any pattern which is specially created and differs in a small way from previous patterns will probably constitute a legally protectable topography.

3.6.4 *Who owns rights in a topography?*

The first step in deciding who owns the rights is to identify the person who created the pattern. This is sometimes a human designer, who will usually create the pattern on screen. Such a designer will usually be an employee, in which case the employing company owns the topography right.

The alternative, which is becoming increasingly more likely, is that the pattern was created by a computer without human invention. This occurs by use of a computer running a program known as a silicon compiler. The circuit specification is fed in and the computer runs the program to work out the best pattern or patterns to provide a chip meeting that specification. This could be the position when a customer supplies a tape of circuit requirements to a manufacturer for a semi-custom chip.

If a silicon compiler is used, the creator of the topography right is the person who made the arrangements necessary for the creation of the topography. Since in legal terms a person can be a corporate body, this almost certainly means that the company owning the computer on which the compiler program is run will be the owner of the topography right, usually the chip manufacturer.

The next step in determining ownership is to ask whether the topography right was generated under a commission – this means one company paying another to carry out the design work. If so, it is the commissioner, the company making the payment, that owns the topography right. This is almost always the case with custom and semi-custom chips and should not pose too many practical problems. The commissioner of a custom chip certainly does not want any other company to have any right to obtain that chip, and ownership of topography right assists this control. Similarly, the commissioner of a semi-custom design will own the topography right in that particular format of the chip; the chip manufacturer who devised the basic layout on which the specific requirements are imposed will retain the topography right in the basic pattern.

The preceding paragraphs assume that the topography was created in the UK by a UK company. For creation by non-UK nationals and non-UK companies, the very existence of topography right in the UK depends on their nationality.

If the creator is a citizen of an EU country or the USA, or if an employing company has its base in an EU country or the USA, then UK topography right exists and that company or person owns the rights.

Similar provisions apply to topographies being exploited by a citizen or a company based in an EU country or the USA, or first exploited in the UK, an EU country or the USA by someone authorised by the owner of the rights in the design who is not a citizen or a company not based appropriately. 'Exploited' in this context means actually sold, advertised or offered for sale. In any of these circumstances, there is UK topography right.

If the creator is a citizen of or based in another country which has not made an agreement with the UK to grant reciprocal protection of topography rights in that country, then UK topography right does not exist at all.

3.6.5 *How long does protection last?*

Protection is automatic in the UK, but this is not necessarily so in other countries (see Section 3.6.8). The topography right comes into existence immediately after it is created and lasts for 10 years from the end of the year in which it was first commercialised. Thus if a topography is created in July 2005 and exploited immediately, the legal right lasts until the end of December 2015. If the topography is never used or exploited commercially in any way, the protection lasts for 15 years from the end of the year in which it was created.

The term is similar to that for design right but for topography there is no equivalent of the licence of right to copy in the last 5 years of its existence. The owner of topography rights has full legal control for the 10 or 15 year period.

3.6.6 *Infringement of a topography right*

The owner has the right to stop others from carrying out a series of acts, most of which involve copying the protected pattern or the protected product in some way. The owner can stop a competitor from copying an electronic record of a pattern, from copying a mask, and from making a semiconductor layer incorporating the pattern. The right extends not just to the whole of the topography, but also to a substantial part of it. Copying a large part of a topography and then introducing a small change into the remaining part does not evade the law.

In addition to copying a protected topography in some way, the right extends to commercial dealings with it, such as selling it, hiring it out or importing it into the UK to sell or hire. This applies even when the chip has been incorporated into other equipment, such as a television set.

It is also infringement to authorise someone else to copy a topography. This protects a chip manufacturer who follows instructions to make a semiconductor product which turns out to be an infringement. The company issuing the instructions will be legally liable.

If a topography owner can show that the topography has been copied, whether as a mask or as a product, legal action can be started. The owner can be awarded an injunction to stop further reproduction, and payment of damages for loss of sales etc.

There is a defence: if the company making the copy can show that it was not aware that a topography right existed, then damages will not be awarded. This is an encouragement to mark the topography although there is no legal requirement to do so. The usual marking is Ⓜ, followed by the owner's name and the year.

If the infringement consists of selling or hiring the chip but not making it, then the only legal remedy is payment of damages which will be set as the equivalent to a reasonable royalty. There can be no injunction to stop further dealing in these circumstances.

3.6.7 Exception to topography rights

An exception exists that allows a competitor to analyse a topography and use the analysis to design another topography. This is known as reverse engineering.

Any competitor is permitted to make a copy of a chip or a mask, but only to a limited extent and for a limited purpose. Any product or pattern record used to produce it can be copied so that the topography, or the concepts or techniques on which it is based, can be analysed or evaluated. Thus the process described above in relation to pirating a chip can be used, and full information about a chip can be obtained in this way. The copying permitted would be in the form of photographing the layers etc. There is no obligation on the owner to provide the information.

Once the topography has been analysed the analysis can be used when another topography is created; if that topography itself is regarded as original and not commonplace, it will be fully protected by its own topography right. Using that right in any way will not be an infringement of the topography right on which it was based through the analysis.

This exception to infringement does not permit straightforward copying of the chip in a production process, there must be the creation of a new right, but the precise wording leaves in doubt whether the earlier topography can be fully copied provided other features are added. If so, the protection of a manufacturer's investment will be much less strong than appears to have been intended.

3.6.8 The position in other countries

The main difference in the laws protecting topography in other countries is that in some cases registration is essential, otherwise legal rights do not exist. In all registration systems, the application for registration must be filed within 2 years of the topography being exploited commercially in any country of the world.

The USA, Japan and all pre-2004 EU countries except the UK, Belgium and Ireland have registration systems. This is an exceptionally active legal area, with many countries introducing laws relating to topography rights. Up to date advice is needed by companies with an interest in this field to protect their export markets in the best way.

Looking more closely at the US law, it was introduced in 1984 and was the first in the world for the reasons mentioned in the introduction to this section. After registration, which must be within 2 years of commercialisation whether in the USA or elsewhere, protection lasts for 10 years from the end of the year of registration or the year of commercialisation, whichever is earlier.

The main difference from UK law is that a topography right applies in the USA only after it has been incorporated into a semiconductor product. After the patterns have been fixed in or on a chip they are protected, irrespective of how they are recorded. A pattern which has not been incorporated into a product is not protected. US protection is therefore narrower in scope than in the UK.

The definition of infringement is similar to UK law in that neither the pattern records nor the chip can be copied and the reverse engineering provisions are also similar. Marking a product is optional, the symbol Ⓜ is to be used, plus the name of the owner of the right. Japanese law is similar to US law.

In no country does topography right extend to the idea or concept on which a topography is based. The concept may be protectable by patent rights if other requirements can be met.

3.7 Overlap among registered designs, design right and copyright

Even though the publicised intention of the UK government in creating design right was to take functional articles out of copyright protection in the UK, there is still some overlap between the two rights, and also with registered designs. The position is summarised below as a series of examples relevant to an engineering environment.

3.7.1 Protection

- If an article is intended for manufacture and the design is new and has an individual character, such as a television set, it can be protected by a registered design, and it will also attract design right provided it is not commonplace.
- If an article could have been protected by a registered design but no application was filed, it will be protected by design right provided it is not commonplace.
- If a drawing is prepared for use in manufacturing an article, such as a television set, the drawing is an artistic work and is therefore copyright.
- If a preliminary drawing is made of an item which will be genuinely artistic, such as a unique statue, the drawing is protected by copyright and the article can be protected by a registered design.

3.7.2 Infringement

The rule states that it is not infringement of design right to do something which infringes copyright. Otherwise, any available right can form the basis of an infringement action, for example, both design right and registered design.

- If a television set is manufactured as an exact copy of either the drawing or the set itself, or with slight variations, there will be infringement of design right and of any registered design.
- If a television set is manufactured without copying but is identical to or very similar to a design registration, the registered design will be infringed. (The design registration right is broader and a longer term, so it is a stronger right.)
- If the drawings of the television set are copied, there is infringement of copyright.
- If a drawing is made of the television set, opinions on interpretation of difficult legal wording differ as to whether there is infringement of copyright or of design right. Since the only commercial intention in an engineering context will be to make an article from the drawing, which is undoubtedly infringement of design right, the practical difficulties are expected to be rare.

3.8 Summary of rights in designs

Almost all manufactured objects which have been designed in the UK will be protected automatically by design right. The owner, usually the employer of the designer, can stop others from copying the product for 5 years, and can claim a royalty payment from copyists for the subsequent 5 years.

The appearance of an article can be registered, provided the shape, pattern, texture etc. is new and has an individual character. A registration can last for up to 25 years provided renewal fees are payable at 5 year intervals. The owner can stop anyone from using this particular design for any product.

Semiconductor chips and the masks used to make them have automatic protection similar to design right, but the right to stop copying lasts for the full 10 years.

Chapter 4
Patents

4.1 Introduction

Every engineer knows the word 'patent' and that patents relate to 'inventions'. What is less well known, sometimes until it is too late, is that a patent is only granted for an invention which is new. This means not known anywhere in the world before an application for a patent is filed.

Disclosing the invention, whether by publishing a description of it, or exhibiting or selling something that incorporates it, means the invention is not new. This applies even if the inventor is the person who does the publishing, exhibiting or selling. The rule 'file before disclosing' is not the sort of rule that can be broken, and many potentially patentable inventions cannot be protected because the inventor did not know this until too late.

There are several other requirements. The invention must not be 'obvious', that is, it must be more than a straightforward application of known principles; it also requires creative content. But the creative level required is much lower than is sometimes thought. Not all patents are related to major breakthroughs in technology, such as the concept of the transistor. A small change to a previous invention can also be patented and most patents are for improvements to an existing product or process. In the case of the transistor, more complex versions of the three-electrode original were designed: improved materials to use as transistors were devised, more reliable ways of attaching metal contacts were developed, and all of these could be patentable inventions.

Patents are expensive, costing very roughly £4000 per country, but they can give broad legal protection over a whole concept and not just a particular device, so the expenditure can be justified.

In the UK, the relevant law is the Patents Act 1977. Other countries have their own laws and there are a number of international agreements which also apply. This chapter looks at the detailed requirements in the UK with some references to the position in other countries.

4.2 How to get a patent

4.2.1 What can be patented?

Many inventions are tangible devices, such as a motor operating on a new principle or an electric cable having a particular construction. Patents are also granted for new materials, such as a new composition of glass for use in optical fibres or a new material for liquid crystal displays. A new method of making something can be patented, such as a cheaper way of making a known insulating high polymer or of doping a semiconductor. Test methods can also be protected.

An overall requirement is that the invention is capable of industrial application. As the examples above will show this phrase is interpreted broadly, it does not relate just to manufacturing industry, but is almost synonymous with having a practical use. For example, it includes agricultural applications and children's toys.

A new way of operating existing equipment can qualify, such as a mode of allocating the routes of telephone calls in a telephone network to minimise exchange overload, or a new way of combining existing measurements from an instrument array resulting in an improved process control system.

4.2.2 What cannot be patented?

A number of things are specifically excluded from patent protection.

4.2.2.1 A discovery, scientific theory or mathematical method

The discovery of a law of nature or a scientific theory about how something works cannot themselves be patented, but a practical application of that discovery or theory may qualify as an invention. So the discovery that $E = mc^2$ is not patentable, but the application of that equation to making an atomic bomb is not excluded from patent protection.

4.2.2.2 Aesthetic creations

Literary, dramatic, musical or artistic works such as a report, a drawing or similar output involving pen and paper or keyboard and screen are protected by copyright (see Chapter 2) and cannot be patented.

4.2.2.3 Mental concepts

A scheme or a rule for performing a mental act, a method of playing a game, or of doing business are not patentable except in the USA, see Section 4.8. These activities take place largely inside the human head. Judging whether a patent was being used would therefore be impossible.

While the rules of playing a game are excluded from patent protection, a board game using a new board and/or new pieces and having new rules, can be patented.

4.2.2.4 Presentation of information

If it is only the information itself which is new a patent is not available, but as with a discovery, if there is a practical use of the presentation there may be an invention.

For example, during the manufacture of a known design of magnetic tape cassette, one guide pole was coloured to assist the assembly process; this was held to be mere presentation of information and a patent was refused. On the other hand, a vehicle speedometer which indicated impact effect by showing both speed and the square of speed, was patented. So was a squash ball which was coloured blue because, surprisingly, a blue squash ball is easier to see. In both cases the information was not merely presented, it was also used in some way. A colour television signal was held to be a physical reality, even though it was transitory, and it was not 'mere presentation'.

4.2.2.5 Computer programs

The precise wording which sets out the exclusion from patent protection refers to 'computer programs as such'. The 'as such' gives a great deal of scope for argument and the meaning is still being tested in the courts.

Many types of computer programs can be patented and Section 4.7 considers how this can be done if all other criteria for patentability are met.

4.2.2.6 Immoral inventions

Inventions which are contrary to public policy or morality are excluded. Thus inventions relating to landmines will not be patented but inventions for detecting or destroying land mines will be processed normally. Methods for cloning human beings or using human embryos for industrial or commercial purposes are excluded.

4.2.2.7 Other exclusions

Inventions relating to methods of treatment for humans or animals and biological processes are excluded, but the boundaries of the exclusion are continually under test by developments in this technical field.

4.2.3 The invention must be new

Patents are only granted for new inventions. The general principle is that legal rights are being granted in return for a disclosure of new information which becomes available for use by others when the patent expires. If that information could be found elsewhere, then the inventor has not kept his or her part of the bargain.

The test for novelty is enormously broad – is the invention publicly available in any way in any part of the world? This applies whether the information is in writing, recorded in some other way or given verbally. It includes information disclosed by demonstration and by selling something which includes the invention.

The test applies to all countries and in all languages. It is not limited to information known to the inventor. So long as the information is available to some member of the public somewhere in the world, under patent law it is no longer new. A description in an unusual language published in an obscure journal invalidates a patent application filed at a later date. One does not have to prove that someone read it. The principle is known as absolute novelty, as distinguished from local novelty when only publications in the country in which the application is filed count against it.

In a court decision, it counted as a disclosure when a chemical composition was sold which included a compound for which a patent was later requested. This was held to be prior use of the invention, although even the manufacturer did not know at the time it was sold that the composition included that compound.

The date the test is to be applied is the date a patent application is filed at the patent office. The application process is described in Section 4.3.

Information disclosed by the inventor or company applying for a patent destroys the novelty of a subsequent patent application just as much as disclosure by a third party. Publication of a research paper or a doctoral thesis which discloses the invention before a patent application is filed invalidates the patent. So does advance advertising which describes the inventive idea.

There are only two exceptions. The first is disclosure which breaches a promise to keep the information confidential; this does not invalidate a patent application provided the application is filed within 6 months of the unauthorised disclosure. The second is if the invention is displayed at certain international exhibitions, but these are few in number. Exhibiting an invention before a patent application is filed almost always invalidates it.

The novelty test is not only very broad, it is difficult to apply. No person or company can be aware of every publication in every language in every country. Sometimes the information does not become available until some years after a patent application has been filed, often at the stage when a patent has been granted and is the subject of legal proceedings. A company allegedly infringing the patent may justify the cost of making detailed searches for publications which pre-date the patent application and therefore invalidate it. The patent owner may not have funded the fullest possible search – an expensive process – and this is a risk which has to be set against the benefits of patent protection.

An invention is regarded as not novel if an earlier publication gives a clear description of something which would infringe a patent claim if it was done after grant of the patent.

4.2.4 The invention must not be obvious

If an invention clears the first hurdle, that of being new, it still has to leap the second by having what is known as an inventive step. This phrase is synonymous with the invention being 'not obvious', and the word 'obvious' is used in a more limited way than in normal language. This has been tested many times in court. To constitute a non-obvious invention, an idea must have a creative element and be more than a straightforward application of known techniques. The background to the level of creativity is all the material which can be used to argue that the idea is not new; starting from that basis one asks if there is a difference, and if the difference is a significant one, and not so small that it is an obvious change.

The next question is, obvious to whom? The addressee is certainly not required to be a top class expert. This was decided in a case relating to the early development of colour television: the invention was the realisation that separating the colour and brightness signals gave important advantages. At that time there were only four research teams in the world working in the field, all of them capable of making substantial technical variations to anything they read. This level of expertise was held to be too high for the test.

The reader of a patent specification is expected to have an appropriate technical background, so that it is not necessary to disclose every tiny detail of the invention, but not to have much imagination. Certainly, this mythical reader need not exercise inventive ingenuity in putting the invention into use. The usual interpretation is that a skilled technician has all the background information, but applies it without ingenuity. This was decided in an early case relating to printed circuits. The patent application disclosed a method of making a printed circuit board (PCB) by covering an insulating base with copper foil, printing the required connections on it with acid-resistant ink and dissolving the remainder of the foil in an acid bath. There was an earlier disclosure of a silk screen process, but this was held not to make the later invention obvious to a skilled technician.

In testing for obviousness, it is essential to consider the date the application was filed and to be aware of what technology was available at that date. Familiarity from long use often makes an invention seem obvious, which may have been the reader's reaction to the PCB example above, but one must always put oneself in the position of a technician working in the field at the appropriate time, which is the date of the patent application.

Would it be obvious to try something? In a case relating to stair lifts for the elderly and disabled, a mechanical seat-levelling mechanism was replaced by an electrical mechanism. This was not an invention; it would be worth trying and the actual implementation would not be difficult.

Often an important factor in a court decision made late in the life of a patent is whether the invention has been a commercial success. If so, there is probably an inventive step. This is especially so if a patent discloses a solution to a problem which has troubled that particular industry for years. The argument, based on hindsight, that the solution is an obvious one and that the patent should not have been granted is unlikely to succeed in the face of commercial success, although clever advertising will be taken into account.

4.3 Filing a patent application

4.3.1 Preparing a patent specification

If an invention seems to be new and not obvious, the first step in getting a patent is to prepare a specification which describes the invention and file it at the patent office, with official forms. The specification must describe the invention in sufficient detail for readers to be able to use the invention. The reader is assumed to have the level of skill which was discussed in Section 4.2.4, that is, to be knowledgeable about the technology but unimaginative in applying it.

Many specifications begin with a description of what is currently known in the technical field, often referring to existing patents, sometimes referring to books or journals. Then a brief reference may be made to a problem which has been encountered, followed by a summary of the invention which solves the problem. Patent specifications for engineering inventions almost invariably contain drawings, because these help to describe how the invention works. The written description of what is shown can be quite detailed, and parts of the drawing referred to are indicated by reference numerals.

If there are several variations of the invention it is only essential to describe one of them in detail, but preferable at least to outline the others.

Since patent specifications are usually written at an early stage in the development of an invention one can never be certain what will turn out to be an important feature, so putting in as much detail as possible gives the basis for expanding on a particular aspect at a later date.

The description of the drawings is followed by the patent claims. These are often difficult to understand because they must meet a stringent requirement. The aim of a main patent claim is to state, in a single sentence, the concept constituting the invention. The claim must include all essential features and, if appropriate, set out how the features interact. The terminology should be as broad as possible, certainly not limited to a particular example, but using general words. There must be no inessential features in the claim.

The main claim is almost always followed by a series of other claims. These may include all features of a main claim by referring back to it, but add additional features. They are known as dependant claims and their purpose is to give fall-back positions. If the concept as defined in the main claim turns out to be known, an added feature may allow a slightly different invention to form the basis of a patent. Each added feature narrows the scope of the claim, because that additional feature is now essential, so usually the more words there are in a patent claim, the more limited it is.

Other claims may define a different aspect of the invention. For example, if the first set of claims covers an article, such as an optical fibre having a particular construction, there may be another set of claims which cover methods of making a fibre that has that construction. If it turns out that the construction itself is not new, possibly the method of making it will still stand up to scrutiny. An abstract is also prepared which summarises the invention in a few lines.

Examples of different types of patent claims are given in Section 4.6.1.

Preparation of a patent specification is an interactive process between the inventor and the patent attorney. The attorney needs to understand the invention (a patent attorney always has a technical background) and it is helpful if the inventor provides a written description as early as possible. Having read it, and possibly viewed any working system, the attorney will ask searching questions about background information to allow the evaluation of novelty and obviousness. Questions will also be raised about possible variations to the invention. The attorney will then draft a specification and send it to the inventor for approval. The inventor should read the draft very carefully to make sure it is absolutely correct and that no errors have been made, whether major or minor. Patent attorneys are used to having their drafts altered; in fact, if no changes are made they are likely to wonder whether the inventor has read the draft at all, much less given it detailed consideration. Both the inventor and the attorney should eventually be satisfied with the final draft.

Any company or individual engineer has the right to prepare and file a patent application and the patent office will provide helpful literature, see www.patent.gov.uk/patent/index.htm. However, it can be a complex business and mistakes can be expensive so most companies use the services of a professional. For information about patent attorneys, see Section 9.2.7.

4.3.2 Establishing a filing date

To start the official patenting process, the specification with appropriate forms is filed at the patent office; there is no fee at this stage. The date the application is received is recorded and this is known as the priority date. The application is also given a number, the first two digits indicating the year of filing. In effect the priority date establishes the position of the invention in the innovatory stream so that it takes precedence over applications filed at a later date. The priority date also starts a timing process and sets dates by which other actions must be taken.

A priority date can be obtained by filing only the description of the invention and the drawings, that is, without claims or abstract. These can be filed up to 12 months later. This gives a very important practical advantage. The invention can be developed and modified and at the 12 month stage another specification can be filed, including all the modifications plus claims which also cover the improved version. This is called a continuation specification. If the invention is still recognisably the same as that described in the first specification, the priority date will still be valid. If parts of the invention are disclosed in the first specification, the priority date will apply to those parts and the relevant claims, while additional material will take the 12 months-later filing date. The aim is that the claims covering the overall concept will take the earlier priority date so that the invention in its broadest form maintains its place in the queue.

If no modifications have been made in the 12 months, one can just add claims and an abstract to the original specification, although it is rare that an engineer is unable to improve on a design in a year. Appropriate forms are required at the 12 month stage and a fee also.

The 12 month limit is very strict. Filing a continuation specification after 12 months plus 1 day is too late; the original application will have lapsed and the priority date will have been lost. The only excluded days are Saturdays, Sundays and public holidays, when the deadline is extended to the next working day. Special arrangements are made during postal strikes. Electronic filing of applications was introduced by the UK Patent Office in August 2004.

The priority-date-plus-12-month deadline is also important when filing patent applications in other countries, but for the present, the procedure in the UK will be followed through.

4.3.3 Searching for earlier patents

The fee required at the 12 month stage includes a fee for the patent office to carry out a search. The object is to test the contents of the specification against the legal requirements that the claimed invention is new and not obvious, by looking for earlier patents for similar inventions.

The patent examiner, who always has a technical training and is a specialist in a particular area of technology, first classifies the invention (as set out in claim 1) in a finely divided technical classification system, then searches the records for existing patents in that technical area. The search covers existing UK patent specifications dated since 1920 (soon to extend back to 1900), and also specifications published by the European and International patent systems (see Sections 4.4.2 and 4.4.3) which came into operation in 1978. US patents since 1971 are included, as are patents from other countries including Japan. Over 45 million patents are included. Examiners may also use papers and technical journals, but because they are not nearly so well classified, these searches are less effective.

The searches are carried out using online database facilities which are continuously upgraded and improved. Newly published patent specifications are added and the database is being extended backwards to include earlier patents. These often provide a surprise: when the Pilkington company applied to patent its float glass manufacturing technique (by which large perfect glass plates are made by pouring molten glass onto the surface of a molten metal), the concept as a whole was found to be covered by patents filed decades earlier, although the invention had never been put into commercial use. Pilkington was able to include details which still allowed a patentable invention to be claimed and on which the success of the process is based. This illustrates the value of having more than one patent claim.

Even in the electrical field early ideas can be surprisingly relevant. The author was once able to show that a granted patent (for an electrical test applied to measure the clotting properties of blood) was invalid, by referring to a medical text book published in 1893 and available only in the British Library.

Once the examiner has located relevant publications having dates earlier than the priority date of the application in question, or failed to find any relevant material as can be the case, the results are listed in a search report. This is sent to the applicant who has the options of altering the application to take account of the cited patents, allowing it to proceed because the disclosures are regarded as substantially different so the invention is still novel and inventive, or abandoning it completely.

It is also possible to have a search carried out without waiting until this stage in the patenting process. For example, a search can take place even before the application is filed and the result used to evaluate novelty and the risk of an obviousness objection. A specification must be drafted to be searched, but after the search it can be redrafted bearing in mind the publications found. Quite often a stronger position can be defined in this way.

Such a search can be carried out by the UK Patent Office, on the same basis as above. Alternatively, it can be done by one of several British commercial or international searching organisations. A further option is for you to use the

European Patent Office (EPO) database, free of charge, by accessing either www.patent.gov.uk and then gb.espacenet.com or directly via www.esp@cenet.com.

4.3.4 Publication of the patent application

Eighteen months after the priority date, the patent application is published by the patent office as a document having a seven digit reference number plus the reference letter A, and is known as the A-specification. At a later date, after grant of a patent if the application survives examination, a second publication takes place using the reference letter B with the same number. The published document contains the specification, claims, drawings and abstract, all as filed by the applicant. It also contains relevant dates, reference numbers and the search report if it is available. An example is given in Figure 4.1 with commentary.

Figure 4.1 is the cover page of an A-specification for UK patent application 2239940. The reference numerals in brackets are used internationally to indicate various items of information as follows:

(12) The publication relates to an application, not a patent.
(19) The application was filed in the British Patent Office.
(11) Publication number.
(21) and (22) Information about this (continuation) application; the first two digits of the application number indicate the year of filing.
(30) The application claims priority for an earlier application, of which the application number and date and country of filing are shown at (31), (32) and (33).
(71) and (72) The applicant and the inventor are different, as is usually the case when the inventor is an employee.
(54) The title is fairly general.
(57) The abstract and the drawing regarded as giving the best indication of the invention.
(43) The specification was published 18 months and 6 days after the priority date at (32).
(51) and (52) Show the classification of the invention in both the international and UK patent classification systems.

Both systems (i.e. the international and UK classification system) have eight broad classes – letters A to H – and each class is divided into groups and subgroups. For the example (Figure 4.1), the classification is:

- International classification F21 V 17/00.
  This is broken down as follows:

F	Mechanical engineering, lighting, heating, weapons, blasting
F21	Lighting
F21 V	Details of lighting devices of general application
F21 V 17/00	Fastening of shades, globes, refractors, reflectors, filters, screens or protective cages.

(12) UK Patent Application (19) GB (11) 2 239 940 (13) A

(43) Date of A publication **17.07.1991**

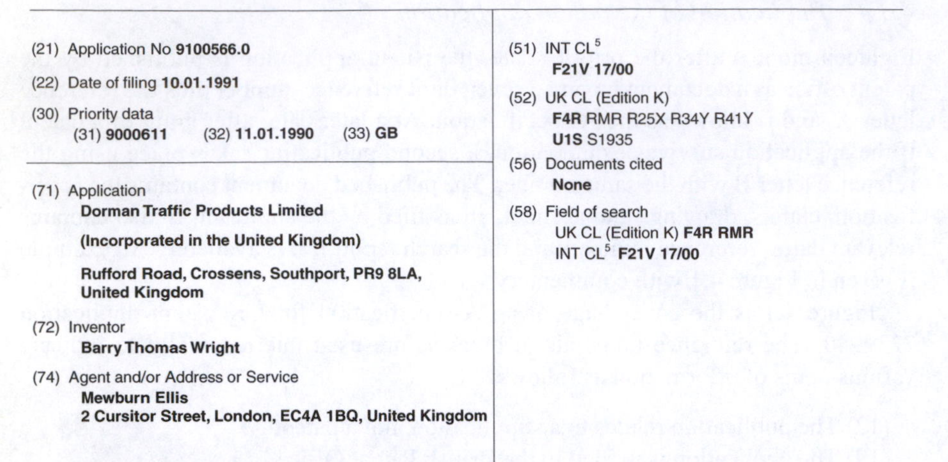

(21) Application No **9100566.0**

(22) Date of filing **10.01.1991**

(30) Priority data
(31) **9000611** (32) **11.01.1990** (33) **GB**

(71) Application
Dorman Traffic Products Limited
(Incorporated in the United Kingdom)

Rufford Road, Crossens, Southport, PR9 8LA, United Kingdom

(72) Inventor
Barry Thomas Wright

(74) Agent and/or Address or Service
Mewburn Ellis
2 Cursitor Street, London, EC4A 1BQ, United Kingdom

(51) INT CL^5
F21V 17/00

(52) UK CL (Edition K)
F4R RMR R25X R34Y R41Y
U1S S1935

(56) Documents cited
None

(58) Field of search
UK CL (Edition K) **F4R RMR**
INT CL^5 **F21V 17/00**

(54) **Warning lamps**

(57) A lamp unit has a connector portion (14) which is a female socket to engage over the upper end of a supporting post (10). A lamp body (12) surmounts the connector portion (14), with at least one lens (15) mounted in the body, and means (42, 44, 46) supports a light source within the space enclosed by the body (12) and lens(es) so as to shine through the lens(es). The support means (42, 44, 46) is a detachable part insertable through the connector portion (14) before the connector portion (14) is engaged over the end of a supporting post (10). Preferred support is a pillar (44, 46) with a light bulb holder (42) at its upper end. The pillar encloses a circuit for causing the light bulb to flash and has a switch (48) for changing between continuous and flashing illumination positioned to project from the pillar into the interior of the connector portion (14).

Fig. 3

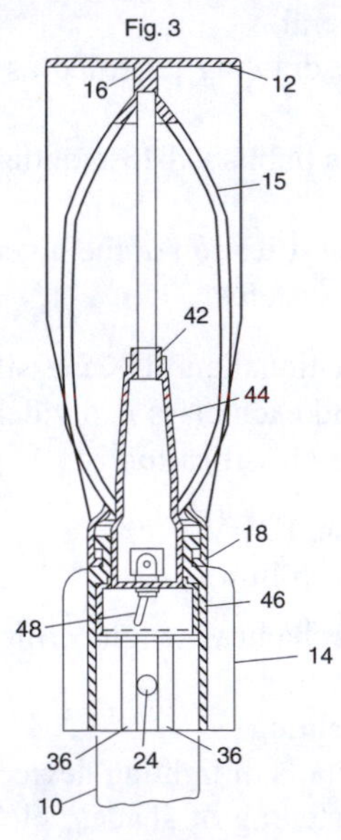

GB 2 239 940 A

Figure 4.1 A UK patent application, front page

- The UK classification uses slightly different groups:
  UK classification HR RMR:

F1	Prime movers, pumps
F2	Machine elements
F3-4	Armaments, lighting, heating, cooling, drying, ventilating
F4R	Lamps etc.
F4R RMR	Lamps other than vehicle exterior lamps.

  The other UK sub-groups listed at (52) are:

F4R R25X	Electric battery lamps (other than for bicycles or vehicles; safety safety lamps, torches and other head lamps)
F4R R34Y	Signalling and warning lamps
F4R R41Y	Light source, incandescent filament lamps.

The examiner clearly decided to search using the international classification and only one of the UK subgroups because the subgroups actually searched are shown beside reference (58). Reference (56) indicates the result of the search; if the examiner had found earlier relevant patents, their numbers would have been given here. In this case, there were no citations.

Until publication of the A-specification 18 months after the priority date, the contents are not available to anyone unless the inventor chooses to disclose them. The only details publicly available are the title, the date of filing and the name of the applicant, sometimes cross-referenced with the name of the inventor. These appear on a card index in the IP Section of the British Library, and in a weekly publication by the patent office called the *Official Journal (Patents)*. Since the title is chosen by the applicant, a company can keep its lines of research reasonably well hidden for 18 months by choosing non-specific words.

Published UK patent specifications are available in most national patent offices around the world. In the UK, 14 cities have libraries providing easy access to patent information, see www.bl.uk/collections/patents/patentsnetork.

Engineers and engineering companies always need to bear in mind the fact that a patent application is published and, therefore, balance the possible disadvantages of this disclosure of the invention to competitors against the advantages of obtaining a patent.

An important right is established from the date of publication. This is the first time a competitor can see what a patent application contains, or the form of the claim the applicant hopes will be granted, so the right to sue for infringement of the patent begins to run from the publication date. The actual proceedings cannot be started until after the patent is granted – after all, the application might fail and it would be unfair to put a competitor to the expense of defending an action before anyone can be certain that a patent right will ultimately come into existence – but any payments for damages can be dated back to the publication date.

The results of the search report are also available at this time, so an interested company can make a judgement as to the likely effect of a granted patent on its business plans, and the likelihood of a patent being granted.

4.3.5 Examination, grant and renewal fees

If the applicant, having received and considered the search report, still wants to proceed with the application, another fee is due 6 months after the publication date. This is the examination fee, paid with a request for substantive examination: a preliminary examination is made when an application is filed to check that all the documents and the appropriate fee are present.

The application is referred to an examiner (often the same one that carried out the search) who reviews the specification and claims, including any amendments made after the search report has been received. A comparison is made with documents cited in the search report and the examiner considers whether the claimed invention is novel and not obvious. The examiner also decides whether the invention has an industrial application and whether it falls within any of the classes excluded from patentability. An objection can also be raised that the application includes more than one invention – this point may have been made at the search stage. The examiner can find that the claims are not clear, not concise, or that they are not supported by the description. Any objections are put into an official letter, and a time set for response to it, usually 6 months.

The applicant for the patent can either provide reasoned arguments refuting the examiner's points or can amend the claims to deal with the arguments, or both. A full response must be provided to all the points raised. The patent attorney dealing with the application almost invariably asks the inventor for assistance at this stage, such as providing or checking counter-arguments, and approving any amendments to the claims.

What is completely forbidden is adding any new examples or explanations to the specification or broadening the claims. All arguments must be based on the material already contained in the specification.

The examiner may issue a second official letter, dealing with the points on which he or she is still not satisfied, and usually gives the applicant 4 months to reply. If there are further rounds of argument, each must be responded to in 2 months, but the whole process must be completed within 4 years and 6 months from the priority date. This sounds a substantial period but it should be remembered that all rights to sue for infringement date back to the publication date so the delays, although sometimes irritating, do not affect the owner's long-term legal position.

Once all the arguments have been met, the applicant is notified that a patent will be granted and, if the applicant is still satisfied that the claims define an invention which is worth pursuing, a grant fee is paid.

The applicant now becomes the proprietor of a patent. The B-specification is published, showing the finally agreed form of the specification, claims and drawings.

The term 'Letters Patent' is not applicable; it related to patents granted under previous patent acts, not to patents granted under the current law, which is the Patents Act 1977.

The proprietor can keep a patent in force for up to 20 years from the date the claims were filed (in the example above) provided annual renewal fees are paid. The first renewal fee is due at the end of the fourth year from filing the claims (in the example above). The fees increase with time and the proprietor has the option of abandoning the patent by failing to pay the fee for any year, but should not do so without careful thought, as the rights cannot be reinstated later if the decision was deliberate.

Since the claims in the example were filed 1 year after the priority date, the total period covered by the patent can be 21 years.

4.3.6 Register of patents

A formal register is kept at the patent office containing information about a patent application and any patent granted on it. Mostly, the register records the names, addresses and dates etc. to be found on the cover page of the specification (see Figure 4.1) but it also records any changes of ownership, and any licences granted under the patent.

Anyone taking assignment of a patent or a licence under it is encouraged to record such a transaction in the register. The penalty for failing to do so is that a later, registered transaction relating to the same patent will be presumed valid. This applies even if it is incompatible with the earlier event, such as a patent being sold twice to different parties.

If the licence is exclusive (one under which the proprietor agrees not to use the invention) the exclusive licensee has an extra reason for registering. An exclusive licensee can sue for infringement in the same way as the owner, but will not be awarded damages by the court unless the licence has been recorded within 6 months of it being granted, or as soon as possible, thereafter.

While the register does not record whether a renewal fee has been paid, this information is available to the public on request. A check can therefore be kept on whether a patent has been allowed to lapse – which is recorded in the register – but remember there is also the possibility of a lapsed patent being reinstated (see Section 4.11).

4.4 Patent applications in other countries

4.4.1 Introduction

This chapter has so far considered the process for patenting an invention in the UK. Almost every country in the world has some form of patent system but each country has its own law and in most cases a separate application has to be filed in each country, dealt with separately through a local patent attorney, and requires payment of official fees and professional fees to the attorney.

Many developed countries have a system roughly similar to that in the UK; other countries make no examination and a patent is granted on payment of a fee with any disputes about novelty etc. being settled in court at a later date.

A patent only has legal effect in the country for which it was granted. If an inventive article is protected only by a UK patent, making the article in China and selling it in the USA are not infringements, although importing such an article into the UK would be. For protection outside the UK, patent applications in other countries are needed. These can be chosen as the countries where major sales are expected, or the countries where major competitors have their manufacturing facilities. For detailed comment, see Chapter 10.

Most significant countries belong to an international convention relating to filing patent applications. When an application has been filed in a 'home' country and a priority date has been established, and if an application is filed within 12 months in another country which is a member of the convention, the second application is treated as if it had been filed in that country on the priority date – the application takes the same place in the queue there as at home. This means that even if a patent application is filed in that other country by a local resident a few months after the priority date in the 'home' country, the convention application in that country still takes precedence, even though it was actually filed there at a later date.

The 12 month time limit is very strict. As with filing a UK continuation application with an updated specification and claims, 12 months plus 1 day is too late and priority has been lost. In the patent system, missing a deadline even by 1 day is fatal to an application.

Two further international agreements make the patenting process outside the UK less daunting than it sounds. The first is a broad system which makes the filing of the application a simple process; the second is the fact that in certain European countries (roughly parallel to the European Community but not quite matching) there is a harmonised system of application and examination. These two systems will now be described.

4.4.2 Patent Cooperation Treaty

Under an agreement called the Patent Cooperation Treaty (PCT), it is possible to file a patent application in the UK Patent Office which has the same effect as filing in any of the countries listed (designated) in that application. The specification can be in the English language. Before the PCT came into effect in 1978, all applications in other countries had to be filed physically in that country and in the language of that country by the 12 month deadline. This made meeting the 12 month convention deadline extremely difficult in some cases, especially in Japan with its time zone far ahead of the UK and the need for translation into Japanese. Convention filings often take place on the last possible day.

The PCT does not apply to all countries of the world, although countries continue to join it. The list of countries applicable at the date of writing this book is given in Table 4.1. Two countries which are not members of the arrangement are Taiwan and Malaysia. If patent protection in these countries is required, direct applications after translation are needed.

The full list of countries in which patent applications are required must be included in the PCT application at the 12 month stage and an appropriate fee paid, depending on the number. If more than 10 countries are designated there is no further fee increase. Countries can be deleted, but not added at a later date, so for important inventions all PCT countries are designated.

Table 4.1 Member Countries of the Patent Cooperation Treaty (as on September 2004)

AL	Albania	GR	Greece
DZ	Algeria	GD	Grenada
AG	Antigua and Barbuda	GN	Guinea
AM	Armenia	GW	Guinea-Bissau
AU	Australia	HU	Hungary
AT	Austria	IS	Iceland
AZ	Azerbaijan	IN	India
BB	Barbados	ID	Indonesia
BY	Belarus	IE	Ireland
BE	Belgium	IL	Israel
BZ	Belize	IT	Italy
BJ	Benin	JP	Japan
BA	Bosnia and Herzegovina	KZ	Kazakhstan
BW	Botswana	KE	Kenya
BR	Brazil	KG	Kyrgyzstan
BG	Bulgaria	LV	Latvia
BF	Burkina Faso	LS	Lesotho
CM	Cameroon	LR	Liberia
CA	Canada	LI	Liechtenstein
CF	Central African Republic	LT	Lithuania
TD	Chad	LU	Luxembourg
CN	China	MG	Madagascar
CO	Colombia	MW	Malawi
CG	Congo	ML	Mali
CR	Costa Rica	MR	Mauritania
CI	Cote d'Ivoire	MX	Mexico
HR	Croatia	MC	Monaco
CU	Cuba	MN	Mongolia
CY	Cyprus	MA	Morocco
CZ	Czech Republic	MZ	Mozambique
KP	Democratic People's Republic of Korea	NL	Netherlands
		NA	Namibia
DK	Denmark	NZ	New Zealand
DM	Dominica	NI	Nicaragua
EC	Ecuador	NE	Niger
EG	Egypt	NO	Norway
GQ	Equatorial Guinea	OM	Oman
EE	Estonia	PG	Papua New Guinea
FI	Finland	PH	Philippines
FR	France	PL	Poland
GA	Gabon	PT	Portugal
GE	Georgia	KR	Republic of Korea
DE	Germany	MD	Republic of Moldova
GH	Ghana	RU	Russian Federation

Table 4.1 (Continued)

LC	Saint Lucia	TJ	Tajikistan
VC	Saint Vincent and the Grenadines	MK	The former Yugoslav Republic of Macedonia
SN	Senegal	TG	Togo
YU	Serbia and Montenegro	TT	Trinidad and Tobago
SC	Seychelles	TN	Tunisia
SL	Sierra Leone	TR	Turkey
SG	Singapore	TM	Turkmenistan
SK	Slovakia	UG	Uganda
SI	Slovenia	UA	Ukraine
ZA	South Africa	AE	United Arab Emirates
ES	Spain	GB	United Kingdom
LK	Sri Lanka	TZ	United Republic of Tanzania
SD	Sudan	US	United States of America
SZ	Swaziland	UZ	Uzbekistan
SE	Sweden	VN	Vietnam
CH	Switzerland	ZM	Zambia
SY	Syrian Arab Republic	ZW	Zimbabwe

PCT applications can be filed in the local language in the patent offices of countries which are members of the convention, but translation into English is required subsequently. The PCT also allows some of the later procedures to take place internationally. An international search can be requested, which can be carried out by one of a number of patent offices having full searching facilities. Such a facility exists at the EPO branch at The Hague in the Netherlands.

For a PCT application, the first round of examination and response can also take place internationally at patent offices in certain countries (but not in the UK, by government decision in the 1970s). The procedure is similar to that for a UK application; the examiner can argue that the invention is not new, or is obvious, in the light of the patents cited in the international search report. The applicant can respond by counter-argument or by amending the claims.

After international search and international examination, a PCT application must be converted into separate applications in the national patent offices of each designated country in which the applicant still wishes to obtain a patent. This is known as the national phase. At this stage the specification must be translated into other languages as appropriate, and a local patent attorney must be appointed in each country. Each application is then dealt with as if it originated in that country; national patent law and its appropriate regulations, fees and deadlines will apply.

At the national phase, 30 months from priority, the financial commitment becomes substantial, but the advantage of the PCT route is that up to this time only one search report has been supplied and only one examiner has been arguing against the application. If applications had been made separately in each designated country, many separate search reports, different examiners and local patent attorneys using different languages would have been involved. The PCT route is more expensive overall, but the expense occurs at a later date in the patenting process, so the applicant can abandon applications for inventions which turn out to be less useful than was first hoped, before major expense has been incurred.

4.4.3 European Patent Convention

Many countries in Europe belong to a European Patent Convention (EPC), which allows patents in these countries to be dealt with centrally up to the stage of grant of a patent. The EPC is not a European Community organisation, it is completely separate. Confusingly, EPC members include most EU countries, and other European countries are in the process of joining. At the time of writing the countries and their status are as shown in Table 4.2.

European patent applications are dealt with by the EPO based in Munich. It is staffed by nationals of many countries and has three working languages English, French and German.

Table 4.2 Member Countries of the EPC

States which are EU Applicants (September 2004) are indicated, and so are members of the EPO but not the EU

AT	Austria
BE	Belgium
BG	Bulgaria (EU applicant)
CH	Switzerland (not a member of the EU)
CY	Cyprus
CZ	Czech Republic
DE	Germany
DK	Denmark
EE	Estonia
ES	Spain
FI	Finland
FR	France
GB	United Kingdom
GR	Hellenic Republic
HU	Hungary
IE	Ireland
IT	Italy
LI	Liechtenstein (not a member of the EU)
LU	Luxembourg
MC	Monaco (not a member of the EU)
NL	Netherlands
PL	Poland
PT	Portugal
RO	Romania (EU applicant)
SE	Sweden
SI	Slovenia
SK	Slovakia (EU applicant)
TR	Turkey

States expected to become members in due course

AL	Albania
HR	Croatia – EU applicant
LT	Lithuania – EU member but not EPO
LV	Latvia – EU member but not EPO
MK	former Yugoslav Republic of Macedonia

Malta is an EU member but not expected to join the EPO

A European application can be filed at the UK Patent Office with all documents in English. Usually it will be filed at the end of the 12 month convention period, to benefit from any updating of the specification since the application established a priority date. Countries of interest must be designated with the application as they can be deleted later but not added. The options are to include the UK in the list, or to keep a UK application separate or to do both for the time being.

The claim requirements for a European application are slightly different from those in the UK. The main claim must have a section known as the classifying portion, which sets out what the technology was before the invention was made, and then what is known as a characterising portion which includes the new features constituting the invention. The main claim as a whole must, as for the UK, have all the essential features of the invention but no inessential features. Also the claims include reference numerals from the drawings to indicate the connection between a general word used in the claim and a specific technical term used in describing the drawings.

When a European application is filed, the procedure is similar to the UK application. A search is carried out and a report on results is sent to the applicant who can then amend the claims or abandon the application. The specification, drawings, claims and search report are published 18 months from the priority date, which may be the date an application was filed in the UK or other national office. An example of the front page of such an application is given in Figure 4.2.

Many of the bracketed reference numerals are the same as those used on the UK A-specification illustrated in Figure 4.1, with the following points of difference:

(19) The symbol of the EPO and its title in its three working languages.
(11) The publication number includes the letter A but the format is different.
(84) The countries designated in the application document are indicated by two-letter reference codes.
(51) The invention falls into two quite different classes: types of magnets and analysis of magnetic properties.

This breaks down as follows:

H01F22	Magnets.
H01F7/22	Electromagnets without armatures having superconducting windings.
G01N24	Investigating or analysing materials by the use of nuclear magnetic resonance, electron paramagnetic resonance or other spin effects.
G01N24/06	Generation, homogenisation or stabilisation of magnetic fields.

The search report was not available at the date of publication (or at least the date the printing was put in hand), but details are available separately.

(72) There were four joint inventors, all named.

If the application is to continue, the search report is used by the European examiner who may argue that the invention is not new, or is obvious or is not the sort of idea which can be patented. The applicant can reply with counter-arguments, or amend

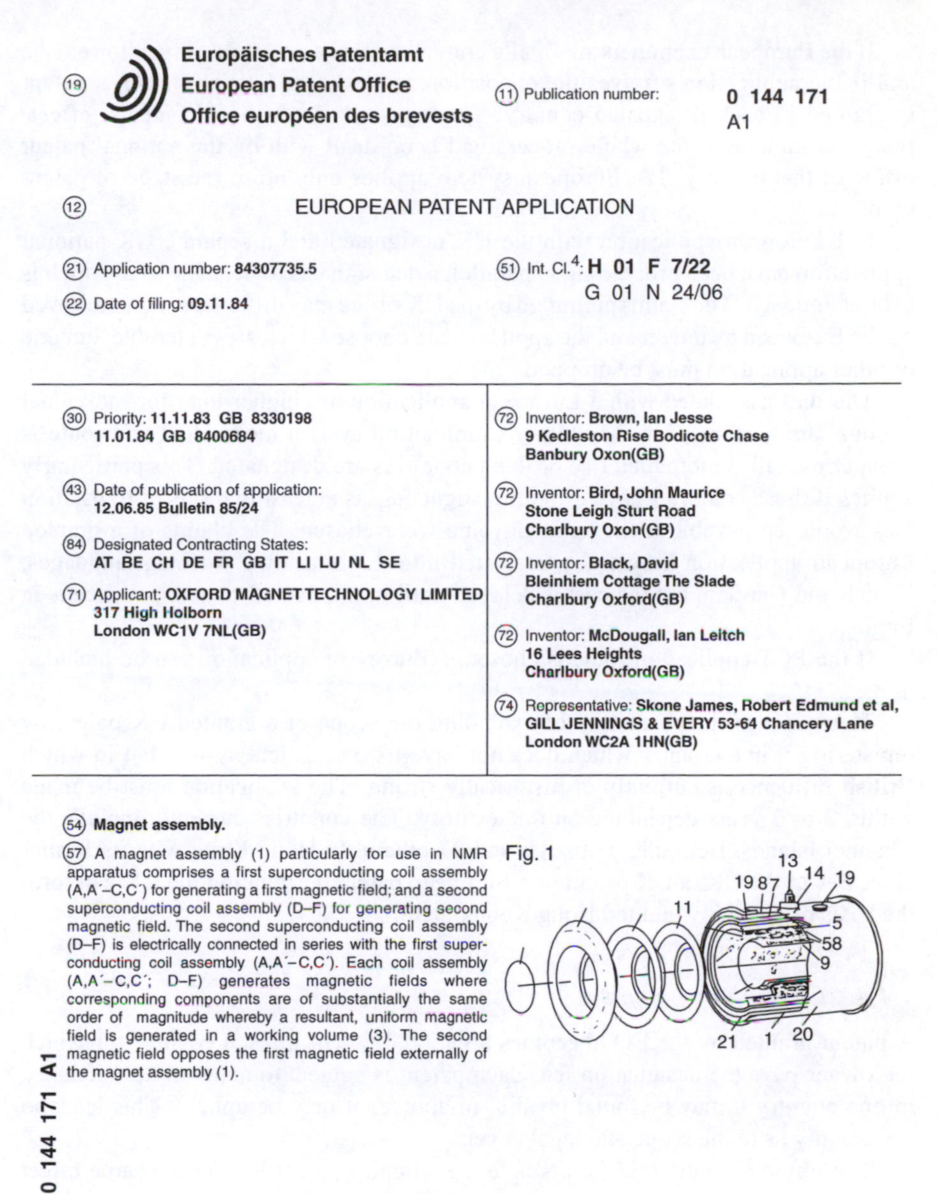

(19) **Europäisches Patentamt**
European Patent Office
Office européen des brevests

(11) Publication number: **0 144 171**
A1

(12) EUROPEAN PATENT APPLICATION

(21) Application number: **84307735.5**

(22) Date of filing: **09.11.84**

(51) Int. Cl.⁴: **H 01 F 7/22**
G 01 N 24/06

(30) Priority: **11.11.83 GB 8330198**
11.01.84 GB 8400684

(43) Date of publication of application:
12.06.85 Bulletin 85/24

(84) Designated Contracting States:
AT BE CH DE FR GB IT LI LU NL SE

(71) Applicant: **OXFORD MAGNET TECHNOLOGY LIMITED**
317 High Holborn
London WC1V 7NL(GB)

(72) Inventor: **Brown, Ian Jesse**
9 Kedleston Rise Bodicote Chase
Banbury Oxon(GB)

(72) Inventor: **Bird, John Maurice**
Stone Leigh Sturt Road
Charlbury Oxon(GB)

(72) Inventor: **Black, David**
Bleinhiem Cottage The Slade
Charlbury Oxford(GB)

(72) Inventor: **McDougall, Ian Leitch**
16 Lees Heights
Charlbury Oxford(GB)

(74) Representative: **Skone James, Robert Edmund et al,**
GILL JENNINGS & EVERY 53-64 Chancery Lane
London WC2A 1HN(GB)

(54) **Magnet assembly.**

(57) A magnet assembly (1) particularly for use in NMR apparatus comprises a first superconducting coil assembly (A,A′–C,C′) for generating a first magnetic field; and a second superconducting coil assembly (D–F) for generating second magnetic field. The second superconducting coil assembly (D–F) is electrically connected in series with the first superconducting coil assembly (A,A′–C,C′). Each coil assembly (A,A′–C,C′; D–F) generates magnetic fields where corresponding components are of substantially the same order of magnitude whereby a resultant, uniform magnetic field is generated in a working volume (3). The second magnetic field opposes the first magnetic field externally of the magnet assembly (1).

Fig. 1

EP 0 144 171 A1

Croydon Printing Company Ltd

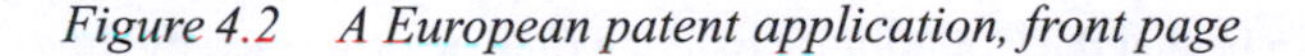

Figure 4.2 A European patent application, front page

the claims or both. There may be second and subsequent rounds of objections and counter-arguments. A European application is dealt with by three examiners working as a team and, if possible, being of different nationality with different natural languages.

If the European examiners are finally convinced that a patentable invention exists, and if the application survives the opposition process (see Section 4.5.4), a patent is granted in each designated country. These patents are national patents, effectively the same as if the whole process had been dealt with by the national patent office of that country. The European system applies only up to the stage of patent grant.

If a European application with the UK designated and a separate UK national application have been proceeding in parallel, a decision has to be made as to which is to be abandoned. The claims permitted by the UK office may differ from those allowed by the European examiner and the applicant can choose which are preferable, but one or other application must be dropped.

The fees associated with a European application are higher than for individual national applications, but the unitary examination system means that this route is cheaper overall if more than two or three countries are designated. This particularly applies if those countries have different languages, as at least two sets of translation fees would be payable if the national route were chosen. The claims of a granted European application have to be translated into the other two working languages, French and German, but the cost is delayed until grant, and all correspondence is in English.

If the PCT application route is chosen, a European application can be included in it.

There is also the possibility of extending the scope of a granted UK patent by registering it in a country which does not have its own patent system but in which British influence is currently or historically strong. The application must be made within 2 or 3 years depending on the territory. The countries currently include the Channel Islands, Bermuda, Gibraltar and 27 others. In Hong Kong with its former close ties to the UK, a UK patent or a European patent designating the UK can form the basis of a quickly granted Hong Kong patent.

4.4.4 Community patents

A patent granted by the EPO becomes a national patent in each country for which the owner pays the designation fee. Each patent is subject to national decisions, so in one country it may be found invalid, in another it may be upheld. This leads to uncertainty as to the scope and legal cover.

The obvious solution is for a single Community patent having the same effect in all countries and a Community court to deal with disputes. The concept has been debated for over 40 years and the differences of law and language were almost resolved in 2004.

The arrangement was to be that the Community patent would be processed by the EPO in English, French or German as usual. In a neat trick, the European Community was to become a member of the EPO, so patent owners could either designate individual countries or designate the whole Community and have a single patent valid in all countries. A Community patent court was to be set-up in (probably) Luxembourg to deal with infringement action and issue injunctions and award damages.

Unfortunately, the long-running differences of opinion about working languages prevailed and it is impossible to guess when or even whether there will ever be such a system.

4.5 Attacks on patents and patent applications

4.5.1 Introduction

Just because a patent has been granted does not mean it is safe from attack by competitor companies. A patent can be the subject of legal action at any time in its life, and objections can be raised by others even before a patent has been granted.

4.5.2 Objections to a UK patent application

As soon as a patent application is published 18 months after the priority date, any competitor or other interested party has access to it and to the search report. Its relevance to a competitor's business can be evaluated. Before the application proceeds to grant, anyone can write to the patent office arguing that a patent should not be granted, giving reasons. For example, this could be done if the competitor knows about a publication not included in the search report which provides grounds for arguing that the invention is not new or is obvious. The interested party can do nothing more than make observations which the examiner will take into account during the examination process; it is not an interactive process.

4.5.3 Objections to a UK granted patent

After a patent has been granted in the UK, anyone has the right to challenge it on the same grounds that could have been raised by the examiner. As with an objection to a patent application, the most likely basis will be knowledge of a publication prior to the priority date which was not included in the search report and therefore was not considered by the examiner. Such an objection is known as an application for revocation of the patent.

Objections can also be raised that the description in the specification is not sufficiently detailed for the invention to be implemented, that is, the skilled technician who is assumed to be reading a specification cannot use the invention without applying inventive ingenuity. This ground could be based on experiments carried out which show that the invention does not work in the way it is alleged to work. The patent office does not carry out tests but confines itself to the description on paper.

A further objection may be that the patent has been granted to the wrong person, if someone else claims to be the true owner of the invention or the true inventor. This objection is only available up to 2 years after the patent was granted: the others can be made at any time during the life of the patent. Any of the objections described above can also be raised in other disputes relating to the patent, the most important being if the owner sues for infringement. The company sued can defend itself by attacking the validity of the patent.

The grant of a patent may therefore be far from the end of the story: a patent is never safe from challenge.

4.5.4 Objections to European patent applications

Using the European route gives the possibility of an attack before the final grant of a patent which is stronger than can be made to a UK application but which is subject to a deadline.

After the European examiner has agreed that all objections have been met, a notice to that effect is placed in an official publication. This sets the start date for a period of 9 months during which anyone can object to a patent being granted on the grounds that the invention is not patentable, does not describe the invention sufficiently, or any of the other grounds described above.

Such an objection is known as an Opposition to the Grant. It is dealt with by a formal procedure in the EPO in Munich. The applicant for the patent can defend the application by counter-arguments or by amending the claims. There may be a mini-trial in Munich called Oral Proceedings during which both sides argue their case, the procedure lasting usually half a day but one day maximum. Only when the European office is fully satisfied that a patentable invention is disclosed will the application be allowed to proceed and patents granted on it in the designated countries. The opposition procedure can take several years to complete.

4.6 Infringement of UK patents

4.6.1 Interpreting patent claims

The general principle which applies after a patent is granted is that its proprietor can stop anyone else from doing something which falls within the scope of any one of the patent claims. The claim or claims must be looked at word by word, phrase by phrase, and compared with what the other party is believed to be doing. The description in the specification, along with the drawings, are used to help understand the claims, but they do not limit them. Comparing the drawing in a patent specification with an allegedly infringing article is not the true test, except for claims in UK specifications which refer to the particular drawing – other countries do not have this type of claim.

The principle of interpreting patent claims will be described using UK patent number 1,487,464, owned by Industrie Pirelli SPA. This patent was granted under the UK law preceding the Patents Act 1977, when some legal provisions were different, but many remain unchanged. The 2 million numbering sequence began when the 1977 Act came into force.

The invention in this patent was made to solve the problem of using optical fibres, which are thin and fragile, in a cable capable of being handled as roughly as one with copper conductors. If the cable is a telecommunications cable, transporting it to a site and laying it in underground ducts can subject it to major stresses. Optical fibres can only be stretched a small amount without breaking, whether the tension is caused by

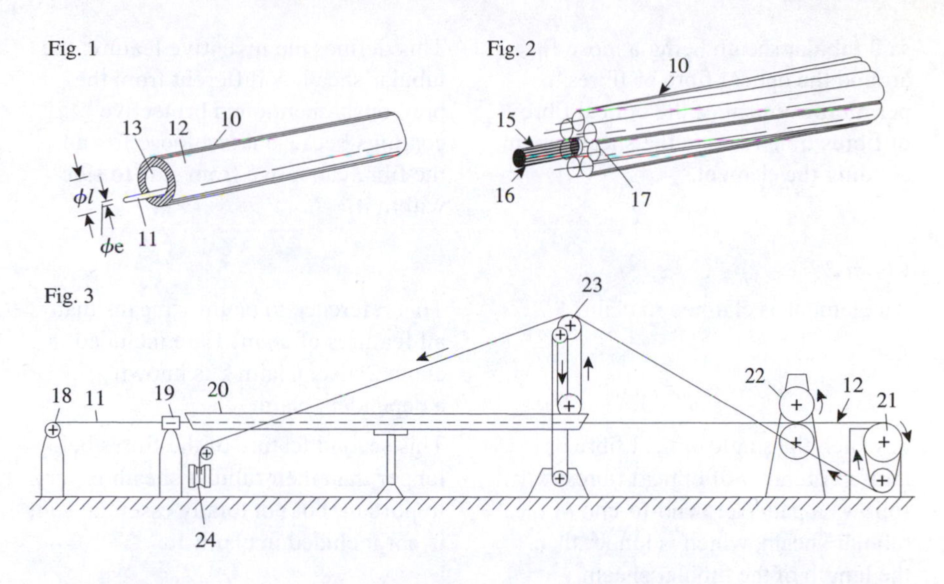

Figure 4.3 Three illustrations from UK patent 1,487,464, Pirelli

stretching or bending of the cable, and there were considerable practical difficulties in using optical fibres for telecom cables.

The solution was to put the optical fibre into a loosely fitting tube. If the tube is bent sharply, the fibre can move within it to maintain a greater radius of curvature. The tube is arranged to be shorter than the fibre so that the fibre takes up a sinusoidal shape within it; the tube can be stretched when the fibre takes up a wave form of greater wavelength, but does not break. Of these two features, the first is more fundamental and therefore is included in claim 1 of the patent.

Figure 4.3 shows the 3 illustrations from UK patent number 1,487,464. Some of the 18 claims, with comments on important features, are listed below.

Claim 1

An optical fibre element for an optical fibre cable . . .	The claim covers just the element and not the whole cable: the element may have uses other than in cables.
comprising a single optical fibre or a plurality of optical fibres laid-up together . . .	The same principle is applied whether there is one optical fibre or more.
the or each optical fibre being bare or enclosed within a protective layer . . .	Some fibres have a protective coating – the claim covers both coated and uncoated fibres.
and a tubular sheath enclosing the single optical fibre or the plurality of optical fibres . . .	Any shape of sheath is covered, so long as it is tubular, e.g. circular, elliptical or square etc. in cross section.

said tubular sheath being a loose fit around the optical fibre or fibres to permit movement of the optical fibre or fibres transverse to the sheath upon bending the element.

This defines the inventive feature; the tubular sheath is different from the previously mentioned protective coatings because it is a loose fit and the fibre can move from side to side within it.

Claim 2

An element as claimed in claim 1 . . .

This reference to claim 1 means that all features of claim 1 are included in claim 2 also. Claim 2 is known as a dependent claim.

in which the single optical fibre or laid-up plurality of optical fibres follows a path from end-to-end of the tubular sheath, which is longer than the length of the tubular sheath.

This second feature of the fibres being longer than their tubular sheath is very important, but not totally essential so it is not included in claim 1.

Claim 3

An element as claimed in claim 1 or claim 2, in which said tubular sheath comprises plastics material.

In claims 1 or 2, the tube could be made of any material. Plastic has properties which make it a preferred material for the tubular sheath but not essential, so claims 1 and 2 are not limited to use of plastic. 'Comprises' means the sheath can be wholly plastic or only partly plastic.

Claim 4

An element as claimed in claim 3, in which said plastics material comprises a thermoplastic material.

Claim 5

An element as claimed in claim 4, in which said thermoplastic material comprises polypropylene or polyethylene.

Further limitations are placed on the material for the sheath.

Claim 6

An element as claimed in claim 3, in which said plastics material comprises an elastomeric material.

This sets out an alternative to plastics material and it therefore refers back to the highest-numbered claim which does not refer to a plastics material.

Claim 7

An optical fibre element substantially as herein described with reference to and as shown in Figure 1 of the accompanying drawings.

The specific reference to Figure 1 means that only an element looking very much like the drawing would infringe the claim. This type of claim is included just in case all the features believed to be new and inventive in claims 1–6 turn out to have been published elsewhere. The intention is that if all else fails, the exact construction might be worth protecting.

Claim 8

An optical fibres cable comprising a plurality of optical fibre elements laid-up together or at least one electrical conductor and at least one optical fibre element laid-up together, the or each optical fibre element being as claimed in any one of claims 1–6.

This claim introduces the idea that the optical fibre element can be part of the cable with an electrical conductor also. The references to any one of claims 1–6 means that any of the versions of the element claimed in them can be used.

Claim 9

An optical fibres cable substantially as herein described with reference to and as shown in Figure 2 of the accompanying drawings.

As with claim 7, possibly the exact construction of cable as illustrated will be a useful basis if claim 8 turns out to be not new.

Claim 10

A method of forming an optical fibre element as claimed in any one of claims 1–6, said method comprising the step of extruding said tubular sheath over said optical fibre, or over said plurality of optical fibres which are laid-up together, after coating said optical fibre or plurality of optical fibres with a composition inhibiting adherence of said fibre or fibres to said sheath.

All previous claims cover an optical fibre element or a cable including such an element, no matter how the element or the cable was made. This claim protects a method of making the element and is known as a method claim. It refers to extrusion of the tubular sheath and to coating fibres with an anti-stick composition. If the sheath was made by a process which does not fall within the term ‘extrusion’, this claim would not cover it.

Claim 17

An optical fibre element as claimed in any one of claims 1–6, or a method as claimed in any one of claims 10–14, in which the or each tubular sheath is circular in section with an inner diameter not less than three times the outer diameter of the bare single optical fibre or of the protective layer on the single optical fibre or of the plurality of optical fibres which are laid-up together.

There has been no reference in the claims so far to the relative sizes of the sheath and the optical fibre within it. Claim 17 introduces this by giving a preferred minimum ratio of the diameters and can be applied to any one of a number of previous claims.

Claim 18

An optical fibre element as claimed in claim 17 or a method as claimed in claim 10, in which the or each tubular sheath inner diameter is between five and ten times said outer diameter.

This gives the preferred range for the relative sizes of the fibre and the sheath, again as a fall back position in case the three times ratio in claim 17 turns out to be not new.

The Pirelli patent was during its life regarded as a master patent. It provided a widely applicable solution to a serious problem. The invention can be implemented in ways not described in detail in the specification or shown in the drawings, but which would still be covered by the broad claims which protect the principle of using a loose sheath.

However, since handling of optical fibres was causing practical problems in the early to mid-1970s for all companies interested in optical fibre telecommunication cables, other solutions were found. The priority date of the Pirelli patent was July 1974. Another patent was filed in October 1974 and was granted to Siemens A.G. as number 1,457,868. The passage of the Siemens patent through the UK Patent Office was quicker, so its seven digit number is lower than the number of the Pirelli patent. The two applications proceeded in parallel and the claims cover different solutions to the same problem, that is, different inventions.

Siemens also use a concept of relative movement of an optical fibre within a containing chamber, but in this case include a series of radial fins which are closed at their radially outward end by a cover and are formed into a helix around a high tensile supporting core. The optical fibres can move within the spaces between the fins and the cover, so that when the whole construction is bent the fibres can maintain maximum radius of curvature at all points. If the supporting core is stretched, the helix pitch increases and the fibres are free to move inwards towards the core.

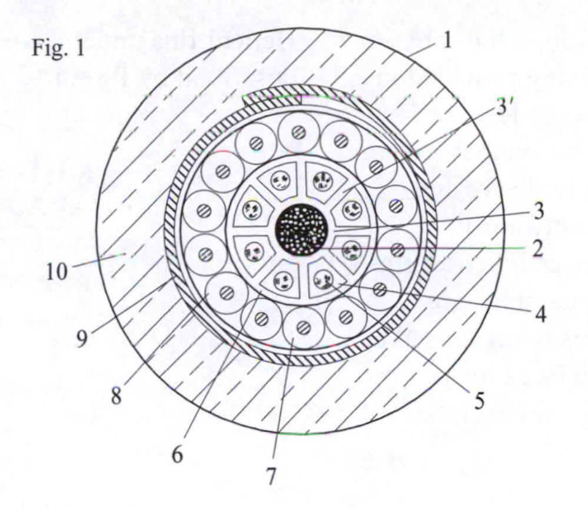

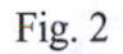

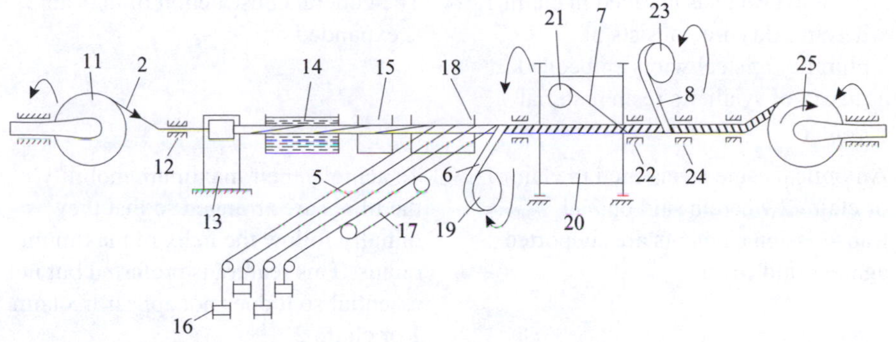

Figure 4.4 Two illustrations from UK patent 1,457,868, Siemens

Figure 4.4 shows the drawings from Siemens patent number 1,457,868, and comments on some of the claims are listed below.

Claim 1

An optical telecommunications cable …	The whole of a cable is being claimed in this construction.
comprising a plurality of elongate optical transmission elements arranged around a high-tensile core …	The number of optical elements is not specified, it is a 'plurality' which means two or more.
each of said optical transmission elements being movably accommodated in a respective one of a plurality of elongate chambers . . .	The movability feature is the same overall concept as in the Pirelli patent but is applied in a different way and was derived independently.

extending helically around the cable axis, said chambers being provided in a chamber member in the form of a layer of thermoplastic synthetic resin material surrounding said core and having a plurality of outwardly extending, integral spaced, helical fins between which said chambers are defined, said chambers being closed at their radially outward faces by a covering surrounding said chamber member.

Helical fins under a cover are not described by Pirelli.

Claim 2

An optical cable as claimed in claim 1, wherein said core consists of a plurality of steel wires embedded in a matrix of synthetic resin material.

The general construction of the cable is expanded on.

Claim 3

An optical cable as claimed in claim 1 or claim 2 wherein said optical transmission elements are supported against said cover.

To achieve their maximum mobility, the fibres are arranged so that they initially follow the helix of maximum radius. This feature is preferred but not essential so it does not appear in claim 1 or claim 2.

Claim 4

An optical cable as claimed in any one of claims 1–3, wherein said chamber member is formed by a profiled strip wound helically about said high-tensile core.

This is the first reference to how the cable is made. A cable made by any technique other than using a profiled strip and winding it helically will not infringe this claim, but could infringe one of the previous claims.

Claim 6

An optical cable as claimed in any one of the preceding claims, wherein said fins have a pitch of 0.2 to 2 m.

This introduces a range of possible helix pitches, to be added to any previous claim in case a construction is found which has all features except the same numerical value.

Claim 11

An optical telecommunications cable substantially as herein described with reference to Figure 1 of the drawings.

The specific reference to the figure means that any cable must look very similar to it or it would otherwise not infringe the claim.

Claim 12

A method of manufacturing an optical telecommunications cable as claimed in claim 1, comprising the steps of rotating said high-tensile core maintained under tension about its own axis: providing said chamber member on said core: placing said optical transmission elements in said chambers without subjecting them to tension: and thereafter applying said covering to said chamber member.	This refers back to claim 1, that is, to the most general claim. The steps of a possible manufacturing method are given and this claim would be infringed only if all steps were taken. A cable made by a different manufacturing process could still infringe claim 1.

The examples of the Pirelli and Siemens patents show that similar inventions can be made independently, especially when there is a practical problem to be solved and several research teams are working on it. Both patents were considered by industry to be master patents.

4.6.2 Patent infringement

A patent is infringed only if every feature within at least one of its claims is used in the allegedly infringing article or process. Referring back to the Siemens patent example, claim 1 would not be infringed by a cable which did not have a high-tensile core, even if it had helical fins, a cover and optical fibres arranged between the fins. In practice, the high-tensile core is essential if the cable is to be usable. If, for some reason, the high-tensile core was made of copper wires, claim 2 would not be infringed because steel wire is claimed in it, but claim 1 would be infringed because copper wires constitute a high-tensile core.

It is therefore necessary to include only essential features in claim 1, and then to add more detail in subsequent claims.

When a patent claim covers an article, such as the construction of an optical fibre cable, it is easy for a patent proprietor to buy a sample and check whether all features of the claim are included. This is less easy to do for method claims such as Pirelli claim 10 and Siemens claim 12. For such a claim, one can buy a sample of the cable and observe if it was probably prepared by the claimed method; with a cable, it is relatively easy to check whether the sheath has been extruded, as required by Pirelli claim 10.

However, if the method claim relates to a method of manufacturing a known material, such as the thermoplastic with which parts of the cable are made, then analysis of the material itself is unlikely to indicate its method of manufacture, and more supposition has to be applied before infringement can be judged. For a claim covering a method of testing, it is even more difficult to collect information on whether a patented method is being used.

But suppose a claim is thought to be infringed, what can the patent owner do?

If a patent claim relates to an article, the proprietor has the legal right to stop anyone from making the article, selling it, using it, keeping it or importing it into the UK. If the claim relates to a new material, the proprietor can control any method of making that material. If the claim relates to a new process for making a known material, the control extends to using that process and to the product made by that process. However, making the known material by a different process, outside the claim, would not be an infringement and the proprietor could not stop such manufacture.

In effect, patent proprietors have a negative right – they have the right to stop anyone doing anything within their patent claims. But use of the invention even by the patent proprietor may turn out to infringe a claim in an earlier patent owned by someone else; owning a patent does not mean you can use the invention yourself. This often comes as a big surprise to a patent owner.

4.6.3 Taking action against infringers

After a patent is granted, legal proceedings can be started for infringement of one or more of its claims. The proceedings can be brought by the proprietor or by an exclusive licensee, either in the High Court or in the Patents County Court or, by agreement of both the patent proprietor and the alleged infringer, in the patent office.

While legal proceedings cannot be started until after the grant, the acts which have taken place since the publication of the application will be infringements, if the patentee wins. The reason is that after publication, the alleged infringer knew about the contents of the specification and is, therefore, liable for infringing activities. (The complex procedure for patent infringement actions will not be described here. Professional advice is essential.)

If an infringement action is started, the allegedly infringing company has the right to challenge the patent, using any arguments about novelty, obviousness, lack of sufficient information etc. as could have been used in objecting to its grant. If the proprietor or the exclusive licensee wins a patent infringement case, they can be awarded payment of damages, an injunction stopping further use or delivery-up of infringing articles to the court, or any combination of these remedies. For further details see Chapter 9.

4.6.4 Threatening to sue

When the owner or exclusive licensee of a patent suspects that it is being infringed, it may seem to be a good idea simply to write to the suspected infringer, threatening to sue if the infringement continues. It may appear that this is an effective way of dealing with the situation and of avoiding the cost of professional advice, but in fact it may initiate extra expenses. In some cases, a company threatened in this way has the right to bring a legal action on the basis that the threats are groundless. This applies if it turns out that no claim of the patent is being infringed, or that a claim is infringed but is invalid.

If the threat is groundless, the company threatened can obtain, from the court, a declaration that the threat was unjustified, an injunction to stop further threats and damages for any loss caused by the threat.

This protection against being threatened applies to secondary infringers, that is, those who trade in a patented product. It does not apply to a primary infringer, that is, anyone who is making or importing a patented product or using a patented process.

What is permissible is to write to suspected infringers, notifying them of the existence of a patent. However, unless the recipient of such a letter takes legal advice, their response is often puzzlement; they may not be aware of the law and may continue their activities. But such a letter must be sent before any more aggressive correspondence, to protect the patent owner's position.

4.6.5 Exceptions to patent infringement

There are some exceptions to patent infringement. The major exception is the use of an invention by the Crown, that is, use on behalf of any government department or the armed services or for foreign defence purposes with the agreement of the British government. If such use occurs, the Crown informs the patent owner about the use, but is not obliged to ask permission. The Crown must pay compensation, but this may occur some time after the event. Furthermore, if the Crown has a record of the invention made internally by a government department (so it does not count as prior publication), the Crown escapes the obligation to pay compensation at all. The use does not have to be by the government department itself: one company can be authorised by the government to use a patent owned by another company, for the purpose of supplying equipment to the Crown.

The second exception is that acts which are done privately and for purposes which are not commercial do not infringe a patent. However, if a patented product is sold to someone for private and non-commercial use, the sale would probably still infringe even if the use did not. Carrying out experiments relating to the invention is also permitted, such as trying to improve it.

4.7 Patents for computer software

Both the UK Patents Act 1977 and the European Patent Convention state that a patent shall not be granted for 'a computer program . . . as such'. Similar exclusions apply in other technically advanced countries. Computer programs are widely protected by copyright, but the much broader protection of patents (where the concept is protected and not merely the precise way it is expressed, as in copyright), means that companies with an interest in software make strenuous efforts to get patent protection for programs.

Many of them succeed, although some fail. There have been many decisions, both in the UK Patent Office and the UK courts at various levels and also by the Board of Appeals of the EPO, and on the whole the approaches are similar. Patents are granted for programs which have a technical character. The technical character can relate to a technical problem or a technical field, but merely running a program on a computer does not constitute a technical character. There must be a useful effect in the world outside the computer, or an improvement in the system which includes a computer running the program, or an improvement to the computer itself.

Looking first at useful effects outside the computer, Vicom Systems applied for a patent through the European route for the digital processing of images by use of a convolution integral. Although the processing was carried out by a computer program operating on numbers, those numbers represented physical entities, that is, a picture which was enhanced by the process, and so a patent was granted. The claim was initially for 'a method of digitally processing data' but this was rejected as too abstract. The granted claim was for 'a method of digitally processing images'. This illustrates the need for careful and expert presentation of patent applications in general, and software-related inventions in particular.

In an application by Lux Traffic, a method of controlling traffic lights by extending the green period when a moving vehicle was detected was patentable, so the technical character does not have to be hi-tech.

In contrast, IBM applied to the European office to patent a program in which input of a linguistic expression caused a semantically related expression to be output. This was rejected as automation of a linguistic system, with no relation to any physical entity.

An application by Hitachi was also rejected: the program constituted a compiler, that is, it translated source code written in a high-level, human comprehensible language into the binary object code with which a computer works. The inventive feature was that scalar instructions in the source code were detected and translated into vector instructions, so that the compiled program could operate on powerful computers capable of running vector instructions and therefore of high-speed operation. The rejection was based on the argument that the program operated on numbers representing merely another computer program and not, as in Vicom, physical entities, that is, improved images.

Turning now to internal improvements in the computer system, IBM requested a patent for procedures applied in a computer network. Connections among a terminal and two or more applications programs were maintained concurrently, and several data files in remote processors could be processed simultaneously. This software implementation was held to be for improved communications within the network and a patent was granted. In an application by Bosch, an improvement to a computer reset procedure was found to be patentable.

The examples above should indicate that consultation with a patent attorney, who is an expert in this field and therefore fully up to date with case law, is especially important for potentially patentable computer software. There is also a strong need to present a software invention in the most advantageous way.

The enormous commercial importance of software means that the limits of the law will continue to be tested as patent attorneys try to extend patent cover. Conversely, some organisations, mainly US-based, argue that software should not be patented at all. Both the UK Patent Office and the EPO set up consultations in the early 2000s and the results were a resounding 'Yes' for patent cover. The resistance still continues with political pressure being applied. The EPO would like to change the definitions so that programs 'on their own' are not patentable and may issue a directive in 2005.

4.8 Patents for business methods

Another exclusion from patentability in the UK and Europe is 'a scheme, rule or method for . . . doing business'. When Merrill Lynch applied to protect a system for trading in market securities by setting values in a computer for 'buy' and 'sell', it was held that there was no technical result. To date, the UK authorities have held firm, but in the USA such inventions are patentable. There are hundreds of business method patent applications pending in the EPO, which may develop a more lenient view.

Until the mid-1990s, there were few applications in this field in any country. Then in the USA a small company, Signature, was granted a patent for a computer-controlled financial system under which investment funds were pooled. This gives the advantage of scale but the fine detail of the changes in value must be recorded for tax purposes and the system allowed this by a 'hub and spoke' arrangement. A much larger firm, State Street Bank, challenged the patent, won in the first court but lost on appeal. The appeal decision sent the message that business methods, if novel and not obvious, were patentable in the USA. At about the same date amazon.com patented its 'one-click' technology for Internet shopping. By allowing shoppers to enter payment and delivery details with one click, the percentage of abandoned orders was markedly reduced. Aggressive approaches for licence fees followed in this and similar cases.

A practical issue for the authorities and for potential users/infringers is that there is little recorded prior art against which to judge novelty. From the point of view of inventors, a US patent application at the very least should be considered, even if the UK and EPO authorities maintain their stance.

4.9 Government security

The UK government has the right to restrict the publication of patent applications relating to the defence of the realm or the safety of the public, if it considers that such publication would be prejudicial.

The system is that the patent office scans each patent specification for indicative words (e.g. bomb, missile) and if such a word is found, notifies the applicant that publication of the information or its communication to a third party in any way is prohibited. The office also notifies the relevant secretary of state.

The list of words to identify inventions considered likely to be a security risk is used by clerical staff, not technical staff. This means that a prohibition order can be attached inadvertently, for example, if one refers to an object for underwater use as 'a submarine device'.

If the contents of the specification relate to atomic energy, the UKAEA has the right to receive the information directly from the patent office to review it. In all other technical areas, before the application reaches the date of 18 months after its priority date at which publication of the application would otherwise take place, the secretary of state can see a specification only if the applicant gives permission. It is advisable to do so as otherwise there is no possibility of the prohibition order being removed quickly.

Most orders are lifted on inspection of the specification by the appropriate government department. If, however, the order is kept in force, the application can proceed to grant through the examination process but cannot be published. The applicant may not apply for a patent through the European or PCT routes, but can usually get permission to apply in individual countries which have appropriate security procedures, for example, the NATO members.

Failure to comply with a secrecy order is a criminal offence, and a director of a company which fails to comply may be personally liable to a fine of £1000 or up to 2 years in prison.

Another restriction is that UK residents are not permitted to file patent applications in other countries unless a UK application has been filed at least 6 weeks earlier. This is the official time it takes to decide if a prohibition order is necessary. Engineers who are normally resident in the UK are strongly advised not to file in another country during a short visit, such as for a conference or a company visit. British nationals who are long-term residents in another country will almost certainly be subject to any secrecy regulations applying locally.

4.10 Patent ownership and transfer of rights

Being an inventor and owning a patent are not necessarily the same thing. The inventor is the person who devises the invention; if two or more engineers generate the idea while working together, then they are joint inventors, provided each has made a genuine contribution to the inventive concept. Merely testing an idea on the instructions of the inventor, or building the first device to the inventor's sketches, or being the manager of the inventor's department, are not sufficient to constitute inventorship. The inventor must be named on the patent application forms and has the right to be named in the granted patent.

If the inventor-engineer is an employee and the invention was made as part of the engineer's normal work, then the engineer's employer will own the rights (see Chapter 7 for full details). If the invention was made during work carried out under contract, the contract conditions may state that the company paying for the work is to own the patent rights, and may even have the right to file the patent application.

After a patent has been granted, it can be sold to someone else – assigned – or someone else can be permitted to use the claimed invention, that is, the patent can be licensed (see Chapter 9). If the owner is willing to grant licences to anyone who applies, the patent can be formally stamped 'Licence of Right' and the renewal fees are then halved. When a patent is not being used in the UK, a third party can apply to the patent office for a compulsory licence by arguing that the invention is not being used for the benefit of the UK, but this procedure is rarely used.

4.11 Renewal fees and failure to pay

The first annual renewal fee for a granted patent is due 3 months before the end of the fourth year from the filing date, and annually thereafter. If any fee is not paid, the patent office sends out a reminder in the 6 weeks after the due date, and the fees can be paid up to 6 months later with an additional fee. If the fee is not paid within the 6 month late-payment period, the patent lapses. If the non-payment is deliberate, for example, if an improved version of the invention is undoubtedly outside the scope of the claim and a decision is made to abandon the protection, then the patent is totally dead.

If the proprietor intended to pay the fee and a mistake of some form was made, then the proprietor can apply to restore the patent. The application must be made within 19 months of the date the fee first fell due, and the proprietor must show that reasonable care had been taken to pay the fee and that the non-payment was beyond the proprietor's control.

This is a strong requirement. Patent owners are expected to set up their own reminder systems (the patent office does not guarantee to provide one, although it almost always does so); leaving payment to inexperienced staff or to a banker's order system without ensuring that the bank complies have both been regarded as not good enough to justify restoration of patent rights.

While a small company or individual may be able to show that care had been taken and that the lapse was inadvertent, a large company with an in-house patent department will find it extremely difficult to satisfy the patent office that sufficient care had been taken to justify restoration of patent rights.

If the arguments are sufficiently convincing, the patent comes into force again, but anyone who has started to use the invention, after the date the late renewal fee could have been paid and before the application to restore is filed, has some rights to continue such use without payment of royalty. Making preparations for such use also acts as a qualification for a right to continued use.

4.12 Marking

Readers will have seen products marked 'patented' or 'patents pending'. There are several rules to apply as far as the marking of products is concerned.

If a manufactured object is patented, or is the subject of a patent application, it is permissible to label the product to that effect. If the object is not itself patented but was produced by a patented process, it can still be referred to as 'patented' or 'patent applied for'. This might be appropriate for a pharmaceutical product which has been known for many years but which is now produced by a patented manufacturing technique. The marking is permissible as long as the article or its method of production are patented or the subject of an application, and for a short period after any patent expires. If, however, there is no patent or application, any person putting on such a false marking can be made to pay a fine.

Labels on products which were correct at the time of manufacture do not have to be removed when the patent expires, so one is not obliged to chase up one's customers over the last 20 years to ask them to remove the labels.

Similar rules apply in other countries. This leads to practical difficulties if an engineering product is exported widely. For each country, any marking should correctly indicate whether a patent or patent application is relevant.

There is a good reason in the UK to put a patent number on a patented product. This is because of the rules applying to the damages payable if a patent is infringed; if the infringer can show that he or she had no reasonable grounds for supposing a patent existed, damages will not be awarded. Merely marking the article 'patented' without including a patent number is regarded as being insufficient warning.

If a patented product is sold in another country having a similar sort of rule, it is highly beneficial to set up a system to make sure that the correct patent number is marked depending on the proposed export destination of the product, or on its sale in the UK. For widely exported goods, this can lead to a heavily bureaucratic system, but it will minimise the risk of legal action.

4.13 Patents and standards

One way of ensuring wide use of an invention is to arrange for it to be essential in the implementation of a standard. This provides almost a captive market but there are a few snags. Some standard development organisations around the world insist on patents being licensed without payment. Also it is not easy to get an invention accepted as essential because other companies in the field will be arguing for use of their own inventions. Some standards committees have closed memberships so even presenting an argument may be close to impossible.

In electrical technology the most likely areas are audio and video coding and telecommunications. For example, use of MPEG-2 for video compression is based on a pool of about 100 patents owned by several companies.

Use of a patent in a standard can be highly profitable. Unisys Corporation has a patent for a date compression algorithm used in association with British Telecom's BTLZ algorithm by many modem manufacturers who pay a one-time licence fee of US$20 000.

Small companies with a strong position can charge high fees. Interdigital Communications Corporation of Philadelphia owns patents for a Time Division Multiple Access invention used in mobile phones. After some argument AT&T took a cross-licence and paid US$2.5 million, other manufacturers of mobiles then paid lump sums of US$20 million each.

The politics of getting into a standard may be difficult, but the rewards can be dramatic.

4.14 Patents in other countries

The major developed countries have systems similar to the UK and the European Patent Offices. When an application is filed a search is made and the invention is examined for novelty, non-obviousness and other factors. However, in many other countries there is no search or examination and a patent is granted on payment of a fee. Its novelty and non-obviousness are tested only if a dispute about it comes to court.

In Japan, the legal definitions allow inventions of smaller scope to be patented than in the UK. This is one of the reasons why Japan has a very high proportion of patent applications for its population size.

The system in the USA is different from all other countries in two important ways. The first is that one can disclose the invention by use or by publication, then file a patent application up to 1 year later; this is known as a grace period. An inventive article can be made and sold so that the market is tested out and a US patent application filed within 12 months of the first disclosure. A valid US patent would still result, but the disclosure would count against an application made in any country having an absolute novelty system, that is, all other major developed countries. Claiming priority under the international convention does not overcome this lack of novelty.

The second difference is that the description of the invention in the specification must be more detailed than is required in the UK. Since the USA is such an important country, patent attorneys usually draft specifications with US requirements in mind, rather than preparing a detailed specification for the USA and a less-detailed version for other countries.There is also an obligation to disclose the best method of implementing the invention that is known to the applicant at the time the specification is being drafted.

4.15 TRIPS and patents

The TRIPS Agreement (TRIPS – Trade-Related aspects of Intellectual Property Rights) places on countries, which have signed it, the obligation to make patents available for product inventions and process inventions in all fields of technology, subject to the normal tests of novelty, inventiveness and industrial applicability. Inventions must be patentable whether the invention was made in that country or elsewhere, and must protect imported products as well as those made locally.

There are three permissible exceptions to patentability:

Inventions which are contrary to *ordre public*.
Diagnostic, therapeutic and surgical methods for the treatment of humans or animals.
Plants or animals other than micro-organisms and also biological processes for the production of plants or animals.

A patent for a product must give the owner the exclusive right to make it, use it, sell it or import it. A patent for a process must cover the process itself and also products obtained directly by the process. Owners have the right to assign patents and to grant licences.

An owner must disclose the invention in a clear and complete way so that a skilled person can carry out the invention, and a country may optionally have a 'best method' disclosure obligation.

The minimum provision is for patent protection to last for at least 20 years from filing the application.

Readers are reminded that different countries will phase in the arrangements at different speeds.

4.16 Utility models

Some countries have both a conventional patent system and a lower level system, known as petty patents, utility models or certificates of utility. In Australia it is known as an innovation patent to distinguish it from a standard patent. In addition, Belgium and Ireland have short-term patents which can be granted quickly.

The protection is more restricted in several senses: it is usually for a shorter term than the conventional 20 years applicable to patents, typically 6 or 10 years; there is no search or examination for novelty so the fees are substantially lower than for patents; while the same requirements for novelty apply as is the case with patent applications, these are only tested in proceedings in court for infringement of the utility model registration.

Utility model protection, as the name implies, is almost always limited to tangible objects, such as an electrical circuit. Neither manufacturing processes nor test methods can be covered in most countries, although this is not always the case.

If a patent application is filed in a country with a utility model system, there is the option of converting it to a utility model application.

The EU is considering the introduction of a Community Utility Model, but the idea is at a very early stage at the date of writing this book.

4.17 Summary of patents

A new article, machine, material or process can be patented, provided a patent application is filed before the invention has been disclosed in any way. A full description of the invention and a definition must be included, forming one or more patent claims. A search is made to check that the invention is new and not obvious, that is, not too similar to previously available information. If the invention meets the legal requirements, a patent is granted which can last for up to 20 years provided annual renewal fees are paid.

The owner can stop any third party from doing any act or making any thing which is covered by the wording of any patent claim.

Chapter 5
Confidential information

5.1 Introduction

Many readers would have been asked at some time in their professional lives to sign a confidentiality agreement. This often happens when visiting another company's development facilities. All engineering companies have information that they would not like to fall into the hands of a competitor, and use of confidentiality agreements is one way of protecting it.

What readers may not realise is that even without signing an agreement they must keep certain types of information confidential. This applies particularly strongly to employed engineers, who have significant obligations to their employers. This chapter describes those obligations, and sets out the background for protecting valuable information.

The scope of the chapter is limited to engineering information in a commercial context. It does not consider government secrets, defence security, or literary or personal confidences.

5.1.1 What is 'confidential information'?

Unlike the other five types of intellectual property right (IPR) covered by this book, there is no statute law which can be consulted for a definition of what is legally protectable. The law of confidentiality has been developed by judges in court as they have considered disputes involving alleged secrets and made decisions on what was equitable and reasonable in the circumstance. This body of case law is the basis for the current position. The relevant cases span 150 years and most of the decisions have been practical and pragmatic, so the position is usually easy to understand once the background has been explained. Many of the cases involve engineering or at least manufacturing companies, so they provide relevant examples.

At one end of the scale are secrets in the normal meaning of the word – facts known only to a very small number of people. Such secrets can be kept only if they cannot be detected when a product is sold. For example, the recipes for Drambuie and Benedictine liqueurs are still secret, because the products are sufficiently complex to make their analysis very difficult, even with today's sophisticated chemical techniques. The hitherto secret recipe for Coca Cola soft drinks was allegedly published in 1993.

In contrast many engineered products are easy to dismantle so that full details of their construction are available. Whether the expense of this is worthwhile depends on the product. Semiconductor chips are often reverse engineered by pirates because they are so expensive to design and so profitable to copy. However, there is no automatic right to use the IPRs protecting an article, just because it is easy to reproduce it. A genuine engineering secret might be, for example, a particular heat treatment applied to a material, because the treatment cannot be detected afterwards. Business secrets are more often commercial in nature, such as the profit on a particular product line or the company's current strategic plan involving, perhaps, a takeover bid or the sale of a subsidiary – information which is so vital that it could affect the share price of a public company if it became known.

Confidentiality can also apply to collections of information, such as detailed test results before analysis. The effort expended in collecting the information can be protected by keeping it secret. A collection can be protected even if each separate item is public knowledge. An example would be a list of customer names and addresses. These would be available from a telephone directory, but the fact that particular people and companies are customers of a firm forms the confidential element. This principle has been confirmed by the courts.

Examples of information which the courts have held worthy of protection include the detailed design of an engine for a moped, a method of constructing swimming pools and a formula for gold ink. Simplicity is not a bar: a simple method of cleaning the hull of a ship merited protection because it saved time and money. The information does not have to be written down or recorded in any way. The expertise in an engineer's head can sometimes be regarded as confidential. Confidentiality can protect the whole range of skill and information which is often referred to as know-how.

5.1.2 What cannot be protected by confidentiality?

Once information is available to the public it cannot be 'made' confidential again. If process details are described in a published paper, one cannot then oblige others to keep them secret. If an incident occurred in public such as a failed public demonstration of a new system, one cannot oblige others to say nothing however embarrassing the event.

Nor can one stop others from generating the same confidential information; coincidences do arise. A century ago, Swan and Edison both developed the electric filament lamp independently, and today's development teams in different companies may produce similar results. Hindering a competing but independent development is not allowed.

Some of the practical steps which a company can take to improve control of its confidential information will be considered in Section 5.6.

5.2 Confidential disclosure

5.2.1 Controlling disclosure of confidential information

The best method of keeping a secret is not to tell anyone, but in an engineering environment it is often essential to disclose sensitive information to other companies or to individual consultants.

The reasons for disclosure are varied. A supplier may need to be told in detail of the proposed use of a component or a potential purchaser may ask for the test results on a new device. In both cases, confidential information will be handed from one company to the other. The disclosing company's position can be protected by a confidentiality agreement, a written record setting out briefly what the information is, who it is being supplied to and the reasons for the supply. Such agreements often include a statement of what may be done with it, for example, the use of information internally for the purposes of evaluation. The agreement should always spell out what may not be done with the information, such as disclosure to a third party or internal use for other purposes, such as manufacture. The company receiving the information signs the agreement, and should keep a copy as a record.

Transfer of confidential information may also occur by observation. A visitor being shown round company premises may observe work in hand, such as a relevant manufacturing technique or even the name of another customer or supplier. A visitor's confidentiality agreement can be used, which the visitor signs as a promise to keep confidential all the information received while on the company's premises. The wording should make it clear that this applies whether or not the information was deliberately displayed. Such an agreement can be used as a matter of routine by receptionists at company premises where sensitive matters are handled. An example of such an agreement is given in Figure 5.1.

The flow of confidential information can be two-way. Two companies may be considering a joint venture (JV) of some sort and will need to disclose their current levels of expertise to each other. Each can then promise to keep received information confidential and to limit its use to evaluation of the potential JV. If it goes ahead, a much more detailed legal agreement will be needed.

The length and level of detail of a confidentiality agreement is very variable. It depends almost entirely on the force with which the disclosing company wishes to make the point that it values its information. When IBM placed a contract on Oxford Instruments plc to develop the first superconducting synchrotron small enough for use in a semiconductor production line, the commercial agreement contained 14 pages relating to disclosure and use of confidential information, and obligations of personnel working on the premises of the other company. This was appropriate for a contract covering a 4 year development programme for a highly innovative x-ray lithography source. The information protected was partly technical, such as the injection energy and the precise shape of the superconducting magnet, and partly commercial.

At the other end of the scale, one paragraph in a letter can be a sufficient reminder of confidentiality, depending on the circumstances.

VISITORS' CONFIDENTIALITY AGREEMENT
TO: Institution of Electrical Engineers
OF: Savoy Place, London WC2R OBL ("IEE")

I.. (insert name)
of.. (insert business address)

acknowledge that during my visit to the premises of IEE, I may be given or have access to confidential information which may be supplied in tangible form or verbally or by demonstration and which may be the property of IEE or of a third party ('the Confidential Information'), and I further acknowledge that the Confidential Information will be made available to me only for the purpose of
....
....

('the Permitted Purpose')

I AGREE:

(a) to use the Confidential Information only for the Permitted Purpose;
(b) to keep the Confidential Information secret and confidential and not to disclose it to any third party except for the Permitted Purpose;
(c) not to copy any Confidential Information wholly or partly except for the Permitted Purpose;
(d) on the written request of IEE to deliver to IEE any tangible items of Confidential Information in my possession;
(e) not to remove any Confidential Information from IEE's premises;
(f) to observe all confidentiality regulations of IEE which may be communicated to me.

I understand that this Agreement does not apply to any item of Confidential Information which comes into the public domain through no fault of mine but that otherwise this Agreement will continue in force without limit of time.

DATED ....200[ ]

SIGNED by..

Full Name ....

Figure 5.1 Example of visitors confidentiality agreement

There are few standard terms relating to such agreements. Usually the wording is negotiated by those involved or their legal representatives. The length of time the agreement lasts should be the period during which the information will be important, plus a safety factor.

5.2.2 Implied confidence

The situations outlined in Section 5.2.1 apply when the company owning confidential information knows that it will, or may, be disclosed to third parties. Sometimes disclosure is quite unplanned. In an enthusiastic discussion between engineers, more may be said than is commercially wise. Sometimes a problem is solved on the spot and a possibly patentable invention is made and disclosed verbally.

Fortunately, implied rights come to the rescue. If it can be argued that from the circumstances of the disclosure it would be obvious to a reasonable person that the disclosure was made in confidence, then the recipient of the information can be legally obliged to keep it confidential.

The first case on this point was heard in 1849 and involved Queen Victoria. She and Prince Albert amused themselves by making drawings and etchings, and they had impressions made to give to their personal friends. A printer was employed to make the impressions, but he also made them up into a collection for favoured customers. Prince Albert sued him and it was held that making the collection was a breach of implied confidence.

Whether English law would have developed the concept of implied confidence quite so strongly without a case involving royalty can only be a speculation, but the principle is now firmly established. It was applied in 1948 in a more conventional case.

Saltman Engineering Company Limited prepared drawings of tools for the manufacture of leather punches and placed a contract on Campbell Engineering Company Limited to make the tools. Campbell produced 5000 punches under the contract, but also made and sold 8700 additional punches. The Court of Appeal decided that, although the contract was silent on confidentiality, it was an implied term that the drawings were confidential documents. This was so even though the punches themselves could now be regarded as public knowledge. Campbell had breached an obligation of confidentiality owed to Saltman.

Even though implied confidentiality exists and can be argued after an inadvertent disclosure, it is highly advisable to record the position in writing. Preferably, a signature should be obtained to confirm that the other party agrees to the implied position. If the other party refuses, then at least one has early warning of a potential conflict, and stronger arguments or legal action can be considered. It is of course far better if the situation is not allowed to arise, with a confidentiality agreement being signed in advance.

In situations such as competitive tenders, where confidential drawings and specifications are sent to several companies, strong conditions of confidence should be imposed. These should make it clear that the information is provided only so that a tender can be made and so that the winner can fulfil the contract. All potential tenderers should also agree to return all documents on failing to win the contract.

5.2.3 Releases from confidentiality

If confidential information is published in some way it can never be retrieved and made secret again. This is a weakness of this type of legal protection and, therefore, particularly careful management practice is needed (see Section 5.6).

Brief reference was made above to parallel technical developments. In such circumstances, one company may decide to keep something secret while the other publishes it or is the first to sell a product which gives away the secret. Alternatively,

RELEASES FROM CONFIDENTIALITY
The obligations imposed by this Agreement shall not apply to:
(a) information which is already in or becomes part of the public domain without fault of the Recipient;
(b) information which can be shown by documentary proof satisfactory to the Discloser to have been in the possession of the Recipient before disclosure by the Discloser;
(c) information which can be shown by documentary proof satisfactory to the Discloser to have been generated independently by the Recipient;
(d) information which has come into the possession of the Recipient from any third party without breach of any confidential relationship

Figure 5.2 Releases from confidentiality

it may turn out that information initially thought to be confidential is in fact public.

To deal with these situations, most confidentiality agreements list the circumstances in which the recipient of the information will be released from the need to keep it secret. Typical release clauses are given in Figure 5.2. If, however, the recipient of the information is the cause of it becoming public, the obligation should still bind that company. This puts it at a disadvantage with others, who will be permitted to use the published information.

5.2.4 Springboard principle

Information received in confidence must not be used in an unfair way after it has become publicly available to some extent. The springboard principle was developed in a case relating to portable prefabricated buildings. The design was ingenious and allowed the rooms to be internally decorated and the building then folded flat for transport.

The buildings were developed by Terrapin Limited and full details were given to Builders Supply Company (Hayes) Limited as part of a manufacturing contract. Some of the buildings were made by Builders Supply for Terrapin, but Builders Supply then began to make the buildings for sale on their own behalf. Terrapin objected, arguing that their confidential information was being misused. Builders Supply pointed out that the details of construction were disclosed in the buildings as sold, so the information was no longer confidential. The decision of the court was that Builders Supply had misused the confidential information. They were not entitled to use it as a 'springboard to dive into the future'. Other companies making such a prefabricated building would have to construct a prototype and make tests. Builders Supply had a long start over such competitors and this was unfair.

This 'springboard' principle has been followed in several subsequent cases, particularly where ex-employees have taken away copies of written information. In some cases they could have used the information without restraint if they had relied on memory, because it was not sufficiently confidential to be a trade secret or the equivalent, so use of it after the employment ended would be unrestrained. Taking a copy seems to bring the springboard principle into effect and to render use of the information unfair.

5.3 Employees

5.3.1 Current employees

Employed engineers have an obligation to protect their employer's confidential information. This is known as a 'duty of fidelity' or 'faithful service', a concept which has grown up over many years. The obligation pre-dates use of contracts of employment and is part of an employee's general duty towards an employer. It applies even if the engineer's employment contract does not refer to confidentiality. An employed engineer must not disclose confidential information which he or she acquires as a result of that employment, whether the information is received from the employer, from a third party as part of the employment duties, or is generated by the engineer during professional activities. Nor should an employee use confidential information for his or her own personal benefit, even if such use does not involve disclosure to others.

Employed engineers must not even put themselves at risk of disclosing their employer's confidential information. In a case in 1946, Mr and Mrs Davis worked for a company called Hivac, which made valves for hearing aids. They wanted to work for a competitor company, Park Royal Scientific Instruments, in their spare time. The court restrained them from doing this. There was no evidence that they would disclose Hivac's confidential information to Park Royal, but there was a serious possibility of conflict of loyalty.

The duty of an employee covers the disclosure of information to an employer as well as maintaining confidentiality. In a case relating to a type of above-ground swimming pool, Mr Bryant was managing director of Cranleigh Precision Engineering Limited, the manufacturer of the pools. He found out about a patent owned by a Swiss company which was potentially damaging to Cranleigh. He did not tell his employer, but left the company and tried to acquire rights under the patent to set-up in competition. The court found he had no right to do so as he should have told Cranleigh that the patent existed.

5.3.2 Ex-employees

In the UK, engineers frequently change employment. Their new job is probably in the same technical field as the old one and they will be required to use their engineering skill and expertise for their new employer. The ex-employer may be concerned that company secrets depart with the person. The English courts vigorously defend the right of an engineer to change jobs within the same technical area, and even to set up in competition with an ex-employer. The legal decisions make it clear that as far as confidential information is concerned the obligations of an ex-employee are at a lower level than for a current employee.

When an engineer changes employer, he or she may continue to use engineering expertise, the sort of thing which has been referred to as 'ordinary skill and knowledge' or an engineer's 'stock in trade' or 'skill and dexterity, manual or mental ability'. What an ex-employee cannot do is use the trade secrets of the previous employer, or other equally important information, to give an unfair competitive advantage to the new employer. One test would be that misuse of the information would be to the serious detriment of the company owning it, but this must be in the view of the court, not merely in the eyes of an irate ex-employer. On the other hand, information can be confidential without being a trade secret or the like, so a current employee would be obliged to protect it while an ex-employee would not. For example, a salesman will know customer names and addresses merely from frequent contact with them and would be restrained from disclosing them while employed, but would be far less restricted after employment ceases on the pragmatic basis that memory cannot be deleted. But supplementing memory by written records is not allowed. In a case where documents were copied and taken away, the court held that this was wrong and that using the copied information to supplement memory was unfair – it acts as a springboard for the new employer. Taking copied papers is of course both infringement of copyright and theft, which may have influenced the decision. The judge said:

> If an employee is sufficiently ill-advised to remove his employer's property when he leaves and that property contains information which was confidential during the course of his employment, he may by reason of his own wrong thereafter be restrained from doing something which, but for his own wrong, he might have been entitled to do.

Sometimes, a departing employee is the subject of a restrictive covenant that is a legal obligation not to work in a defined way. A particular technical field might be excluded, or a named company, or (particularly for salespersons) a specified geographical area. Such covenants usually form part of a contract of employment so that a new employee signs the agreement on arrival at a company. This is clearly not the best time to start negotiating on conditions applying when one leaves the company, and employees often put their signatures to very onerous conditions.

The English courts have taken a very strong line. Such covenants must be reasonable. Engineers must have the right to work, using their general skill. Covenants which might be considered acceptable would be those which apply for a limited period of time – possibly with an ex-employer paying the equivalent of salary. A geographical restraint must be reasonable in the circumstances, such as limited to the close vicinity of a business or to the previous sales territory, or to specified customers. As so often in this area of confidential information, it depends on the circumstances. The courts will protect an employee's right to work, while recognising that a company needs to protect its trading position.

One way in which the courts encourage employers to write reasonable covenants is by rejecting the whole of an agreement or the whole of a clause, if they find one part of it to be unreasonably restrictive. A solution to this is to write a series of restrictions of increasing severity, with the hope that at least the weakest condition will be found reasonable. In general, restrictive covenants should be applied only to very important employees.

5.3.3 Implied obligations of ex-employees

Three clear but limited principles apply to ex-employees:

- An ex-employee must not use or disclose trade secrets of a previous employer.
- An ex-employee must not carry away copies of records for future use.
- An ex-employee can be subject to reasonable restrictive covenants.

But this is not the end of the story. An ex-employee can still be limited by implied terms, even if no restrictive covenant has been signed and no trade secret is involved, but only if valuable company information has been properly managed.

This question was considered in the Court of Appeal in 1985. While the case was not engineering based – it concerned chickens – it gives practical illustrations of several management issues.

The dispute arose between the company Faccenda Chicken Limited and one of its ex-employees, Mr Fowler.

Faccenda's business was breeding and slaughtering chickens, and selling them chilled but not frozen. Mr Fowler had been the sales manager and had the idea of using refrigerated vans for direct supply to supermarkets and butchers. Over several years he built up a profitable operation for Faccenda based on ten vans, each driven by a salesman round a particular area of the Midlands. The salesmen-drivers quickly learned the customer names and addresses, the quantities and qualities required by them, the days and times of day each customer required a delivery, the prices charged to each, and the best routes to follow.

Mr Fowler and Faccenda then parted company. Mr Fowler set-up a rival operation, in the same areas of the Midlands, using refrigerated vans to supply many of the same customers, often at lower prices. He advertised for staff, and eight people, five of them salesmen-drivers, left Faccenda and joined him.

Faccenda sued Mr Fowler, arguing breach of confidence by misuse of the sales information about routes, particularly the information about prices charged. Faccenda lost in the High Court, appealed, and lost again.

One factor which was considered was the attitude of the employer. The sales information, including the information about prices, was generally known, not only to the salesman on a particular route, but to other salesmen and to the secretaries in the company. The management had never given instructions that the sales information was confidential, and had not kept the pricing information, which was now argued to be particularly important, separate from other details.

The court set tests to help in deciding whether an ex-employee is bound by implied obligations of confidence. Factors include:

1. The nature of the employment: if confidential material is handled frequently, there will be a greater obligation of confidence after employment ends. The employee will realise the sensitive nature of the material, in contrast to an employee who rarely handles confidential material. In the Faccenda case, none of the van drivers normally handled confidential information.
2. The nature of the information: some information, while not classifiable as a trade secret, is so highly confidential that it requires the same protection as a trade

secret. If information falls into that category, an ex-employee can be prevented from disclosing it. In the Faccenda case, the pricing information was held to be of such a nature that it could have had the same protection as a trade secret if it had been properly handled by the company. But Faccenda did not take any steps to protect the information.

3. Whether the employer impressed on the employee that the information was confidential.
4. Whether the information was separable from other information.

As mentioned above, handling information in a way which indicates its value will help to place greater obligations on an ex-employee. Faccenda did not issue warnings and did not keep the pricing information separate.

The case still does not provide a definition of 'trade secret' but it gives some useful practical guidance on handling secrets. The principles have been approved and reapplied in many subsequent cases.

5.3.4 Employees and notice period

The preceding paragraphs will make it clear that the obligations of an ex-employee are at a lower level than the obligations of a current employee to an employer.

When an employed engineer gives notice of an intention to leave a company, there is still a duty of good faith to the employer, but it may be less than if notice has not been given. If an engineer starts to set up a competing business in the notice period, under a court decision this is quite permissible. What is not allowed are attempts to entice customers to a new company or making copies of useful records to take away.

For managers, it may be helpful to review with the departing engineer what confidential information the engineer has had access to or has generated, and try to define what belongs to the employee and what to the employer. This will not be easy, but a reminder of the ground rules should be beneficial to both sides.

5.4 Confidential computer programs

Most computer software used by engineering companies comes in the form of commercially available packages, such as CAD programs, spreadsheets and databases. The copies are supplied as object code, that is, in the binary language by which a computer is operated. If a particular program malfunctions, either another copy is easily available, a software maintenance company will modify it, or an alternative package with similar functions can be supplied.

If software is specially written for a company, only the software developer will have the knowledge or expertise to cure a fault. This is because only the developer is familiar with the source code from which the object code is derived. Source code is the program written in a high-level language which, while it is highly specialised, is comprehensible to a human being.

Software developers protect their expertise by keeping their source code closely confidential.

If the program is important to the user company, and especially if the developer is a small firm whose financial future is not secure, the user may want to ensure that the program can still be maintained even if the small firm vanishes.

The usual compromise between confidentiality and security of the user is to set up an escrow arrangement. A copy of the source code is placed with a trustworthy third party who keeps it confidential until a certain event occurs, such as the developer becoming insolvent, when the copy is released to the user who may then study it and maintain the program when a fault arises.

Advice on escrow schemes can be obtained from the National Computing Centre website at www.ncc.co.uk under Services. Sometimes a bank or a solicitor is asked to hold the escrow copy – this is acceptable provided conditions which are physically suitable for storage of software can be ensured.

A practical concern is that the user company cannot be sure that the information placed in escrow is in fact sufficient to permit maintenance of the program. If there is serious concern, then an independent expert can be retained to examine it under strong conditions of confidentiality.

5.5 Unwanted confidences

Many engineering companies, especially large ones with a high public profile, are offered 'good ideas' by members of the public wishing to be helpful. Most of these suggestions are technically impractical or impossible, but occasionally a suggestion is genuinely useful. Frequently, the ideas are offered only on condition that they are received in confidence.

The risk for a company is that the genuinely useful idea is already being developed in-house, but that the person making the suggestion does not believe this and accuses the company of breach of confidence. Alternatively, the proposal is sometimes merely a 'wish list', long recognised by the industry as being desirable, but barred by a technical problem. If the company later comes up with a practical solution, again the individual making the suggestion may accuse the company of stealing the idea.

One solution is to accept an idea from outsiders only if there is no obligation of confidence. This is likely to discourage most offers. An alternative is to agree to look at suggestions only if a patent application has been filed first; the application acts as a record of the idea and bears an official date-stamp. But the individual must invest either time in preparing his or her own patent specification, or must pay substantial professional fees. Contribution of ideas is again discouraged.

Neither of these approaches provides an answer if the letter offering the idea contains a description of it. If a company employee reads the letter, the information has been transferred into the company and the ground laid for arguments of breach of confidence.

One way round this is to have all letters which look as if they might be offering ideas to be opened by non-technical staff. The company can then argue that whoever read the letter did not understand the technical content, so the company has not received the information.

The policy for dealing with such unsolicited technical suggestions depends on whether the company wants to discourage them by strongly worded refusals to accept the information in confidence, or to encourage them by politely pointing out the position and suggesting that the writer re-submits the idea in a way acceptable to the company.

5.6 Managing confidential information

The ultimate test to apply to information is 'would a judge agree that this information deserves legal protection by confidentiality?'. The judge will not be convinced of this if the company claiming to own it did not treat it as secret until too late. A thread running through many legal decisions is that the information must be managed appropriately within a company if it is to benefit from the protection of the law of confidence. A list of practical steps which can be taken is given below.

Marking. Papers and computer records which are regarded as important can be marked, for example, 'commercial in confidence'. If necessary, access can be restricted, for example, to named people. Copies can be numbered and recipients noted. A company should devise a policy on handling such documents – possibly they should be locked away each night, or not left on a desk in an unoccupied office. Obviously, not all papers need to be treated in this way. The inconvenience would be unjustifiable, and the message of treating some information with special care would be diluted or lost.

Separating important information. If information is of great importance it should be kept separate from other information. Documents which are generally available should not include confidential items.

Reminders to employees. Although employees have an obligation to their employers to keep company information confidential, it is prudent to remind them of this. There are several ways of doing so: a clause in contracts of employment is one way, although these are rarely referred to after one arrives in a company; issuing occasional reminders to staff is another. As with marking of papers, there needs to be a balance between keeping the issue in the employees' consciousness and diluting the message by too-frequent reminders.

Different levels of staff within a company will need to be reminded at different frequencies. A chief engineer will be handling company secrets on a regular basis, while young support staff will be less aware of their obligations, so a reminder to them associated with a particularly sensitive project may be beneficial.

5.7 Know-how and show-how

Know-how, the application of technology to a practical situation, can often be valuable but it certainly need not be novel, and may only be protectable by confidentiality. It is defined in an EU regulation as 'secret, substantial and identified in any appropriate form' which has the implication that there must be a record. Some types of information can only be demonstrated, and the word 'show-how' is sometimes used for them.

Even though know-how and show-how are difficult to define, both can be confidential. They are often transferred from one company to another, for example, as part of a licence agreement, by visits of engineers to the recipient of the information to give technical training and demonstrations. Such visits can also be covered by confidentiality agreements.

5.8 Legal remedies

If an engineering company's confidential information is disclosed in some way, either by publication, demonstration or transfer to another company by an ex-employee, it is rarely possible to make the information confidential again, but some legal action can be taken. If the parties cannot negotiate a settlement, formal legal proceedings can be started. There are three requirements:

The information must have been confidential.
The information must have been disclosed in circumstances which impose an obligation on the receiving party to keep it confidential.
The obligation must have been breached.

If all three requirements can be demonstrated, the usual legal remedies are available. These include payment of damages and an injunction to stop further disclosure or further misuse. Legal remedies are described in more detail in Chapter 9.

A disadvantage of confidentiality as a form of protection is that third parties who receive the information innocently, without any notice or implication that it is confidential, cannot be stopped from making further use of it. However, the party which has caused the breach of confidence can be restrained from further use and is thus put at a disadvantage.

5.9 Confidentiality in other countries

5.9.1 General

The decisions of the English courts often extend to former British Commonwealth countries, so similar principles may apply. In continental Europe, the law is civil law, and the code easy to find. In the USA, the applicable law is state law, not federal law; in many states the early English presence still shows, but particularly in the south, the French influence means that civil law is used.

In other parts of the world the existence of a law of confidentiality, and even local comprehension of the concept, varies widely. Unlike other IPRs there are no direct international conventions relating to confidential information, but the situation is being changed by TRIPS (Trade-Related aspects of Intellectual Property Rights).

5.9.2 TRIPS

Under the TRIPS Agreement, undisclosed information – trade secrets and know-how – must be legally protectable, on condition that it is secret, that it has commercial value by reason of the secrecy, and that reasonable steps have been taken to keep it secret. A person lawfully in control of such information must have the possibility of preventing it from being disclosed to or used by others without consent, in a manner contrary to honest commercial practice; this will cover breach of contract and breach of confidence.

Local advice in every country is particularly important for this area of law.

5.10 Summary of confidentiality

Any type of information which is not available, or not easily available, from another source can be protected by confidentiality. If valuable information is to be disclosed to a third party the best practice is for the recipient to sign a confidentiality agreement in advance, agreeing not to disclose or misuse it.

Employees have an implied obligation to protect confidential information belonging to their employer, but the obligation is lower when the employee leaves the company. The way information is handled in a company can affect the implied obligation on employees to treat it as confidential.

Chapter 6
Trade marks

6.1 Introduction to trade marks

The aim of a trade mark is to encourage purchasers to buy your company's product or your company's service by using the mark when they place an order. It follows that trade marks should be easy to remember and sufficiently different from marks used by competitors to avoid confusion.

Trade marks are most important for consumer goods or services supplied to the public, but they can also be used for high-technology products or in heavy engineering. This chapter describes briefly the principles and, it is hoped, the benefits of good trade marks for engineering businesses.

In the UK, trade marks need not be registered, but there are many advantages in having a registration, so this chapter considers first the requirements of registrability. The principles are sensible ones, so they should be used when selecting a trade mark even if there is no plan to register it.

6.2 Registrable trade marks

A trade mark can be 'any sign capable of being represented graphically'. Most marks are words or two-dimensional symbols, or a combination of the two, but it is possible to have marks which are three dimensional, or consist of a sound or a smell.

There are two basic requirements. The first is that the mark must be distinctive, that is, it must be capable of indicating goods or services originating from a particular source and of distinguishing those products from the products of competitors. The second requirement is that the mark does not indicate the type, quality, purpose or geographical origin of the goods or otherwise describe the goods or service, and it is not a word which the trade is likely to use in normal business. Such a word would be taken to mean a product from any source, and not one from a particular company. Additionally, the mark must not be deceptive in any way. If the name of the goods is included in the mark, then others cannot be stopped from using that name, thus 'Coca-Cola' and 'Pepsi-Cola' are both registered for cola drinks.

When a trade mark is a word, it is commonly one of two basic types; either a normal word taken out of context or an 'invented' word. Any word mark can be used alone or with a symbol.

Registration can be handled by the company wanting the trade mark or by a professional trade mark agent, who is often also a patent attorney. A professional is not only familiar with UK law and able to give initial advice on the possibility of registering a particular mark, but is also generally familiar with trade mark law in Europe and other countries, and capable of giving early warning of difficulties abroad if exports are planned.

6.2.1 Types of trade marks

6.2.1.1 Invented words

The concept of an 'invented word' is taken from early trade mark law and not the current Trade Marks Act 1994, but the idea is a good one and is frequently used in generating new marks.

An invented word is a word which has been made up specially for use as a trade mark, a word which has no meaning in the English language. Preferably, the word is short and pronounceable. Good examples are 'Xerox' for photocopiers, 'Kodak' for cameras and 'Dulux' for paints.

When new, such trade marks are often disliked by marketing departments, for the very reason that they do not relate to the goods or product. However, the wide reputation of the three examples just given should show readers how powerful this type of mark can be after advertising and use have generated a reputation in the word.

The word must meet the test of being distinctive. 'Miniclaw' for gardening implements was refused as not being distinctive, and so was 'TCS' for sheet metal, because it is a common abbreviation for terne coated steel.

Further, the word must not indicate the nature of the goods, so 'XpressLink' for telecommunications equipment was refused. The word 'Orlwoola', if used on 100 per cent woolen goods would also indicate the nature of the goods, while if the goods were not 100 per cent woolen it would be deceptive, which is not allowed either, so the word loses the right to registration both ways.

For intended trade in countries where the language is not English it is essential to check the meaning of the invented word in the local language so as to avoid offence or embarrassment.

6.2.1.2 Words used out of context

These are normal English words but used in a different way. Good examples are 'Apple' for computers and 'Jaguar' for cars.

Again, a very important requirement is that the word should not directly refer to a characteristic of the goods or service it is being used to promote. 'Supreme' for photography products was refused and so was 'Optima' for patient membrane oxygenators; both words imply a high quality. But a little ingenuity can help;

'P.R.E.P.A.R.E' for revision guides was accepted; it did not describe the goods because it did not consist exclusively of the word 'prepare'.

The word must be out of context. Calling a telephone apparatus 'One Touch View' describes one of its features and is unregistrable.

6.2.1.3 Symbols and shapes

A trade mark can be in the form of a shape or a picture, either alone or combined with a word or words. While the legal term is 'device marks', they are commonly referred to as logos, although strictly that name applies only to a combination of words and symbols.

The symbol can be simple in style, such as the silhouette of an apple with a bite out of it as used for 'Apple' computers. Alternatively, the symbol can be extremely complex and include pictures of flowers, animals or whatever the owner thinks appropriate. The symbol can be varied over time as graphic fashions change, as can be seen in Figure 6.1, showing how the 'Shell' trade mark for petroleum products has changed over the decades while retaining the same concept.

Symbols without an associated word are excellent for visual recognition (as with ties indicating club membership), but they are impossible to use in speech – think of the former British Rail symbol. They are often used in association with a name of the product or the name of the company.

A symbol which is to be used as a trade mark should not be a shape which results from the nature of the goods or services, such as the shape of a lemon or a banana for trade marks for those fruits, or a spanner for garage services. Other businesses should be allowed to use those shapes if they wish to do so.

Nor should it be a shape which is necessary to achieve a technical result; readers may be familiar with the Philips' rotary shaver with three heads, one at each corner of an equilateral triangle. Philips tried to register this arrangement as a trade mark but failed because the shape is functional. The fact that the heads could be arranged as an isosceles triangle or along a straight or curved line did not overcome the objection.

Most device marks are two-dimensional, but three-dimensional marks are allowed, provided the shape is not determined by the product or its packaging. A good example is the shape of the bottles used by the Coca Cola company. The bottle can be graphically represented by line drawings or by photographs.

6.2.1.4 Other permitted types of mark

A trade mark can consist of a single colour, but combinations of colours are more likely to be distinctive. However, registration of a particular shade of green by British Petroleum for the livery of petrol service stations was the basis of a successful infringement action in Northern Ireland. It is advisable to specify colours by Pantone numbers or the like.

A mark can be a smell. The essential graphical representation consists of an appropriate word, such as 'the smell of fresh cut grass' in an application in which the goods were tennis balls.

Sound can also be registered, for example, by using musical notation.

Figure 6.1 *The 'Shell' trade mark at several dates*

Such esoteric marks are probably more likely to be used by businesses close to consumers and mass marketing techniques than by engineering companies.

6.2.1.5 Choices to avoid

In addition to meeting the requirements of distinctiveness and avoiding marks which indicate a characteristic of the goods or services in question, a new mark must be different from existing marks. The reason is to avoid confusion in the market place. A mark will not be registered if it is identical to, or similar to, an existing mark which protects identical or similar goods or services.

The word 'similar' can extend quite a long way. If a mark is registered for wine, the same or similar mark cannot be registered for whisky by a different owner, because consumers would assume there was only one trade origin of the two products.

Marks will not be registered if they include national flags or coats of arms in a misleading way, or if the mark is contrary to UK or European Community law.

6.2.2 Using a trade mark

When a mark relates to goods, it can be used on the goods themselves, either as a label attached to the article or, if appropriate, embossed on or moulded into the goods themselves. The mark can be used on packaging, on invoices or in advertisements for the goods. For small items, the trade mark can be on a card to which the goods are attached for sale. For services, the mark can be used in adverts and on order forms at or near the place where the service is made available or where the service is actually performed.

6.2.3 Classification of goods and services

Since the object of a trade mark is to indicate the goods or services of one firm as opposed to another, the risk of confusion is usually limited to those particular goods or services.

For trade mark registration purposes, goods are divided into 34 different classes, and there are also 8 classes of services. Those most likely to be of interest to engineers are depicted in Figure 6.2. The division of the classes has been agreed internationally and is very widely applicable.

The classes have a practical benefit: since there is only a finite number of distinctive words and symbols available, use of identical marks in different classes is recognised as not causing a problem. Readers will be familiar with 'Penguin' books and 'Penguin' biscuits and are unlikely to confuse the two.

Existing registered marks can be searched free of charge on the UK Trade Mark Registry website at www.patent.gov.uk:tm.

6.2.4 How to register a trade mark

Registration of trade marks is handled by the Trade Marks Registry, a part of the UK Patent Office. The application must be filed by someone who is already trading or who intends to trade in the goods or supply of the services in question. Usually trade

Classes of engineering goods

Class 6	Common metals and their alloys; metal building materials; transportable buildings of metal; materials of metal for railway tracks; non-electric cables and wires of common metal; ironmongery, small items of metal hardware; pipes and tubes of metal; safes; goods of common metal not included in other classes; ores.
Class 7	Machines and machine tools; motors (except for land vehicles); machine coupling and belting (except for land vehicles); agricultural implements; incubators for eggs.
Class 8	Hand tools and implements (hand operated); cutlery; side arms; razors.
Class 9	Scientific, nautical, surveying, electric, photographic, cinematographic, optical, weighing, measuring, signalling, checking (supervision), life-saving and teaching apparatus and instruments; apparatus for recording, transmission or reproduction of sound or images; magnetic data carriers, recording disks; automatic vending machines and mechanisms for coin-operated apparatus; cash registers, calculating machines, data processing equipment and computers; fire-extinguishing apparatus.
Class 11	Apparatus for lighting, heating, steam generating, cooking, refrigerating, drying, ventilating, water supply and sanitary purposes.
Class 12	Vehicles; apparatus for locomotion by land, air or water.

Classes of engineering services

Class 37	Construction and repair.
Class 38	Telecommunication.
Class 39	Transport; packaging and storage of goods; travel arrangement. Also in this class is distribution of electricity.
Class 42	Scientific and technological services and research and design relating thereto; industrial analysis and research services; design and development of computer hardware and software; legal services.

Figure 6.2 Trade mark classes of goods and services

marks are associated with the manufacturer of the goods, but marks can indicate a different type of connection, such as a retailer selling 'own brand' products, or a licensee of a trade mark owner.

The applicant fills in a form giving an address in the UK, either of the company, or in the case of overseas applicants an address for service, that is, for correspondence with the Trade Marks Registry. The application should also include the required 'graphical representation' of the trade mark. If this is a word without limitation on the script, the word is simply typed on the application form and the word will be protected in any script or font. For words in special scripts or for device marks, the application form must show exactly what is to be registered, such as line drawings from several angles of a three-dimensional mark.

The applicant must list the goods or services for which the mark is to be used. If the same mark is to be used on goods which fall into two or more classes then both or all need to be listed. It is usual to specify the goods in some detail. This makes it easier to bring an action relating to what is specifically listed rather than having to argue similarity to the registration.

The specification should be broad, but not too broad. If a word is to be used for voltmeters it should be registered for 'electrical measuring instruments', but if the specification refers to 'all goods in Class 9', the intention to trade in this huge variety of products would have to be justified and even the largest manufacturer or trader would find that difficult.

For readers wishing to file their own application, the forms and helpful information are available on www.patent.gov.tm. The application is filed at the Trade Marks Registry with the application fee (currently £200 for the first class and £50 for each additional class). The application is date stamped and given an official number. Within a few months, a search is carried out to check for conflicting marks. The search is quite wide. It covers current registrations in the relevant class of goods or services, and also applications for marks not yet registered and marks which have previously been refused registration. It includes marks filed through the Community Trade Mark Office (see Section 6.6.3) and international marks that are validated in the UK (Section 6.6.3).

Sometimes, the application is accepted and registered, but often the trade mark examiner writes to the applicant setting out objections to the registration. These are usually along the lines mentioned in Section 6.2.1, that is, that the mark is not distinctive, that it is likely to cause confusion or that it is identical or very similar to an existing registered mark or an application filed at an earlier date. If objections are raised, the applicant has 6 months to reply. This can either be by correspondence or by asking for a hearing, a sort of mini-trial in which the applicant argues the case before a senior trade mark examiner, known as a hearing officer. Arguments would be that the mark really is distinctive, is not likely to cause confusion or is not similar to an existing registration, with supporting detail. A written decision is given shortly afterwards.

A frequently used way of overcoming an objection that the mark is not distinctive is to file evidence to show that the mark has been used for some time and is in fact well known and does indicate the applicant's goods or services. Usually, someone in the company making the application files a Statutory Declaration, setting out the history of use of the mark, attaching examples of advertisements, and giving details of advertising expenditure. If use has been sufficient, for example, for several years, the examiner will often allow the application to proceed.

For example, the word 'Premier' when applied to luggage is inherently not capable of distinguishing, because it indicates a high quality. The applicant was able to show the use of the mark since 1985, and so the word had acquired distinctiveness and it is registered.

If the applicant's arguments are accepted, or if the application goes through without objection, the mark is advertised in a publication known as the *Official Journal (Trade Marks)*. Within 3 months of the advertisement, anyone can object to the registration, using any of the grounds of argument previously available plus some new ones. For example, a trade competitor might have been using the same or similar mark for some time without registration so that if the application succeeded there would be confusion with an established unregistered mark and a newly registered mark in the same field.

If there is no opposition, the mark is registered. The registration takes the same number as the application, and a Certificate of Registration is issued. The initial registration lasts for 10 years from the application date and must then be renewed by paying a fee which at the time of writing is £200. The registration can then be renewed indefinitely at 10 year intervals. In general, it costs about £600 to register a mark in the UK, when using a professional trade mark agent and encountering no substantial objections, and takes 1 or 2 years to reach registration.

One of the first trade marks to be registered in the UK, in 1876, was the red triangle used for beer by the Bass company, and it is still registered.

6.2.5 Infringement of a registered trade mark

After registration the owner has the legal right to stop anyone using the registered mark in relation to the goods or services specified in the registration. The owner can also stop the use of the registered mark on similar goods or services, or the use of a similar mark on the registered or similar goods or services, provided the public is likely to be confused.

For example, the word 'Viagra' is registered in Class 5 for pharmaceutical products. The owner brought a trade mark infringement action against a company that intended to market a drink under the name 'Viagrene' and to advertise the drink as stimulating the libido in men and women. This was use of a similar mark on similar goods, confusion was likely, and the trade mark owner won.

Use must be in the UK. Use on goods manufactured in the UK for export counts as infringement and so does import of such goods, subject to EU requirements for the free movement of goods (see Chapter 8). In certain circumstances customs and excise officers can assist a trade mark owner by seizing infringing goods as they enter the UK.

Oral use is infringement, and the owner can control any sort of written or printed use, and can even stop labels or packaging bearing the trade mark from being printed if they are not for use in the UK on genuine goods.

However, the goods must be 'similar'. Putting 'Kodak' on a tee shirt did not count as infringement.

A trade mark owner cannot stop the use of a person's own name or address, or the use of geographical or other indications of characteristics of the goods or service. The use of one registered trade mark in a proper manner is not an infringement of another registered trade mark.

Similarly, a trade mark owner cannot stop what is known as 'parallel' or 'grey' imports; when the goods have been put on the market in another country legitimately, the owner cannot prevent them from being imported into another country and resold.

Use of trade marks for spare parts and accessories needs special care. Independent traders are allowed to supply such items in competition with the manufacturer of the original article, but the way these are advertised must be closely controlled. Thus, it is permissible for any trader to advertise 'film for Kodak cameras', but the term 'Kodak film' is an infringement unless it is used by the Kodak company.

If a trade mark owner believes a registered trade mark is being infringed, the first step should be to consider a polite letter asking the other party to stop. Sometimes, the user will not be aware of the registration and nothing more is needed. If a letter does not work, or if the owner is sure the other company is already fully aware of the position, action can be started in the High Court. If the owner acts quickly after detecting a possible infringement, for example, within days, the courts can be persuaded to act quickly also. Defences to the action are that the registration was invalid, or there has been no infringement or the alleged infringer has been using an unregistered mark for some time and has therefore established the right to continued use.

If the trade mark owner wins the infringement action, the court may award an injunction against further misuse of the mark and erasure of the mark from the goods or, if that is not possible, destruction of the goods. These are the most important remedies, but damages and accounts of profits can also be awarded. For further information see Chapter 9. The rights extend not just to the goods, but also to packaging and advertising material, and also to articles specifically designed to make copies of the mark, such as a printing plate.

6.2.6 Comparative advertising

Companies often wish to use the registered trade marks of other companies in advertisements where two products are compared. This is permissible provided such use is honest.

The companies Vodafone and Orange both run mobile phone networks. Orange advertised that 'On average, Orange users save £20 every month', by comparison with Vodafone's equivalent tariff. Vodafone objected, but the court held that the advert was not misleading.

6.2.7 Keeping a registration in force

After a mark is registered, the owner still needs to take great care over how it is used. If the use is careless, or there is no use at all, the owner risks the registration being struck off the register, or amended in some way.

6.2.7.1 Proper use of trade marks

For a proprietor to maintain full rights, a registered trade mark must never be used as a noun or a verb, but always as an adjective qualifying the name of the goods or service. For example, one must refer to a 'Xerox photocopier' and not to 'a Xerox'. This gives a clear indication of the owner's brand of a type of goods. When the registered mark is a word, using it in quotation marks, in capitals, with a capital first letter, or with the symbols ™ or ® or (RTM) all give emphasis to the special nature of the word. This type of use should occur in all written materials, not just in advertisements or labels, but also in company reports, even internal ones, and in business letters.

For a device mark, it is also advisable to ensure strict compliance with the form in which it is registered. This avoids any unauthorised variations being made which might detract from the value of the registration. If the registration is limited to a specific colour, the use of a Pantone number, for example, is highly recommended.

Sometimes, a trade mark is indicated as ® or RTM – this should only occur if the mark really is registered; it is a criminal offence in the UK to claim that a mark is registered when it is not. This applies in other countries also, and for goods traded internationally such a marking must be very closely controlled on a country-by-country basis. In these days of wide international trade, a common practice is to use ® on the goods in all countries, provided the mark in question is registered in the home country.

6.2.7.2 Generic names

If a word mark is used as a noun, there is a substantial risk that it will slip into common use as the name of the goods themselves, not the name of the proprietor's version of these goods. If this happens, all trade mark rights are lost, anyone can use the name and the advertising expenditure of the proprietor has been wasted. Word marks which have become generic include aspirin, nylon and formica. These were all originally trade marks, but came into general use as the name of the goods.

As far as maintaining protection is concerned, it is use of the mark by the trade and not by the public that counts. If the general public starts to use a registered trade mark to describe the goods whatever the source, the proprietor must still make it clear to wholesalers and retailers that the word is a registered trade mark which will be protected by legal action if necessary. The proprietor is then in the happy position of having the public think they are asking for any brand of goods, while suppliers know they must supply the trade marked variety. This can only apply if the mark was registered and used properly as a trade mark before the generic use began, and it is a dangerous situation which can easily slip out of control.

6.2.7.3 Cancellation of registration

If a mark is registered but not used for 5 years, or if it was registered without a genuine intention to use it, then any person aggrieved (usually someone wanting to use that mark) can apply for the registration to be cancelled on the grounds of non-use. To avoid this, a proprietor must make real commercial use on a substantial scale within 5 years. Token, small scale use will not save the registration, but use of a trade mark by a licensee counts as use by the proprietor and will avoid cancellation.

6.2.7.4 Amendment of the registration

If a registered trade mark is not used on the full range of goods or services in the specification the register may be rectified by deleting the unused parts. This is the only type of amendment which is allowed, other than changing the owner's address.

6.2.8 Other types of trade mark registrations

In addition to the normal types of trade mark registration described above, there are two further types; certification trade marks and collective trade marks, which have different objects from normal registrations.

Certification trade marks indicate a particular property of the goods, such as their quality, accuracy or geographical origin. Examples are the kite mark indicating approval of manufactured goods by the British Standards Institution, and the mark 'Harris Tweed', which indicates tweed cloth originating from the Island of Harris. The Institute of Engineering Design also has a certification mark.

This type of mark must be administered by a trade association or something similar that does not itself trade in the goods. The association must supply a set of rules to the Trade Mark Registry which specify the standard etc. with which goods bearing the mark must comply. Any company or person complying with the rules must be permitted to use the mark.

Collective trade marks indicate that the user of the mark is a member of, for example, a trade association, such as the Federation of Engine Re-manufacturers and the International Federation of Inspection Agencies, which both own a collective mark. The mark can indicate geographical origin. The association must supply rules to the Trade Mark Registry which specify, for example, the conditions for membership.

If a trade mark has a reputation in the UK, as a 'famous mark', and someone uses that mark, or a similar mark, even on goods or services which are quite different to those to which the reputation attaches, then there is infringement.

6.3 Unregistered trade marks, 'get-up' and 'passing-off'

6.3.1 Unregistered trade marks

Using a trade mark in the UK without registering it is not only permitted, it generates a valuable right as the mark's reputation is built up by advertising, and by sale of the goods or provision of the services. This applies both when the mark could have been registered but the owner decided not to register it, and when the mark is unregistrable because it does not meet the requirements of the Trade Marks Act 1994. In both cases, use of the mark generates what is known as goodwill.

There is the additional benefit that a mark which in its early days is unregistrable may, after sufficient use, become registrable because its reputation can be used as proof that the mark has become distinctive for the goods or service for which it has been used.

6.3.2 Get-up

Unregistered rights extend well beyond the word or device to which a registered trade mark is limited. The term 'get-up' includes the colour of packaging and the type of lettering used on a label. It includes a slogan used for broadcast or printed advertising. Get-up includes the visual impression of business premises, such as the style of fascia board and the colour of shop fittings. In other words, reputation can extend to all the peripherals around the goods or service that can in any way be associated with trading.

6.3.3 Passing-off

If a mark or get-up has been used long enough to have generated a reputation, the owner can stop a competitor from 'passing-off' goods or services as if they originated with the owner by use of identical or similar marks or get-ups. To do this the owner must show in a High Court action that

- there is a reputation in the mark or get-up
- there has been confusion in the market place
- the owner has suffered damage.

6.3.3.1 Proof of reputation

The owner needs to prove that the unregistered mark or the get-up is distinctive and is taken to show a trade connection with a particular business. This does not mean that the whole general public makes the connection, just the relevant section of it. Usually, a number of people such as wholesalers are asked to make Statutory Declarations, saying that they know the mark or get-up. There is no set period of time required to generate a reputation, the more extensive the advertising, the faster the recognition.

As with registered trade marks, there is no need to prove that those knowing the mark or get-up can identify the owner; the aim of a trade mark is to generate 'same again' business by use of the mark itself, so precise knowledge of the owner is irrelevant.

6.3.3.2 Confusion

Confusion can be caused by a competitor using an identical or very similar trade mark or get-up. It is not necessary for the whole of the mark or get-up to have been used, use of the most memorable part will be sufficient. Again, Statutory Declarations are made by people who have been confused by use of the second mark or get-up. There is no need to show malicious intention.

6.3.3.3 Damage

The owner need not show actual loss of business, such as loss of orders to the competitor, but must demonstrate at least a high risk of loss of business arising from the confusion. In any case, the very existence of confusion puts the owner's goodwill at risk. Damage is not limited to cases where the goods supplied are of an inferior quality to those produced by the owner.

6.3.4 Examples of passing-off

6.3.4.1 Model numbers

In the engineering field, it is quite often the practice to use a company name plus a model number or a type number as a trade mark. While most model numbers are not by themselves sufficiently well-known to form the basis of a passing-off action without use of the company name also, a few very well-known model numbers can reach that level.

For example, the company Hymac Limited made mechanical excavators and in one machine the capacity of the excavation bucket was five-eighths of a cubic yard. It was marketed as the 580, in a series of models known as 580 B, 580 BT, 580 CTS etc. which were very successful. Even when the bucket capacity was increased to six-eighths, the name 580 was still used. Another company began to sell very similar mechanical excavators under the name Mustang 580. They were unable to relate '580' to any physical property of their product and were restrained from using this indication in future. The use of 'Mustang' was not affected.

6.3.4.2 Secondhand goods

Sale of secondhand goods as if they were new can be passing-off, as the manufacturer's obligations to repair or service the goods may differ after the initial sale. This applies especially in the car trade. As soon as a motor car is registered and driven away, it becomes 'not new' and sale of cars even with very low mileage should indicate whether they are genuinely new or secondhand.

6.3.4.3 Business names

The name of a business almost always generates a reputation which can be protected from misuse by others. This also applies to the name of a professional association which is protected, as are the initials used to indicate membership of that association.

The protection does not apply to descriptive names because that would give an unfair monopoly in the words. Small differences between two descriptive names will be sufficient to avoid a passing-off action. For example, the names 'Office Cleaning Services' and 'Office Cleaning Association' were held by the court to be sufficiently different names for there to be no confusion and no passing-off.

It is quite possible for a registered trade mark and an identical registered company name to exist with different owners. The registrar of trade marks and the companies house registrar do not search each other's registers; it is left to the applicants to make their own checks. Companies House will register any company whose name is not identical to one already registered, and introduction or omission of a hyphen, or running two words together, is sometimes sufficient difference. The Trade Marks Office website at www.patent.gov.uk.tm has a helpful link to the Companies House website.

6.3.4.4 Books, journals etc.

The title of a book or a periodical is not protected by copyright and therefore passing-off is the only possible way of stopping misuse. Usually, exact copying is necessary. The magazine titles '*Rubber & Plastics Weekly*' and '*Rubber & Plastics Age*' were held to be not so similar as to cause confusion. For newspapers, the names '*Morning Star*' and '*The Star*' were considered not to conflict. Changing the name of a play from '*Sealed Orders*' to '*Orders Under Seal*' avoided conflict in the theatrical area.

6.3.4.5 Franchising

One area in which licensed use of get-up as well as trade marks is common in the consumer field is known as franchising. In chains, such as Benetton shops selling clothing and Wimpy fast food outlets, each establishment is independently owned, but the style of premises and the quality of the goods sold, and even the level of training of staff, are controlled by the franchisor. The franchisees benefit from central advertising and the wide reputation, and pay a fee for the rights granted to them.

6.3.5 *Remedies for passing-off*

The remedies available are an injunction against further use of the name or get-up, delivery-up of the offending goods for destruction or obliteration of the name etc., and payment of damages or an account of profits. For further details see Chapter 9.

6.3.6 *Trade libel*

The wrongdoings which can be dealt with by a passing-off action shade into other areas of law, particularly that known as a trade libel. This relates to written or oral false statements, whether about goods or a business. An example might be a statement that the goods are worthless. The target of such statements can bring an action for trade libel if the statement was made with malice, but the details are outside the scope of this book.

Consumers are also protected by laws against misleading advertisements and false indications of the origin of goods and services; again these are outside the scope of this book.

6.4 Criminal provisions and counterfeiting

Trade in counterfeit goods, that is, goods bearing a trade mark and purporting to be genuine, is well known in the area of luxury goods, such as 'Rolex' watches or 'Lacoste' sportswear, and in high-value goods, such as pharmaceuticals. It can also occur in the engineering field, and supply of counterfeit aircraft parts and vehicle brake pads is known. Usually the counterfeits are of inferior quality, sometimes to the extent of being highly dangerous.

Often the problem is international. There are major differences in national laws, especially in the Far East, which can only be dealt with by international lobbying on a long-term basis. In the UK, there are criminal provisions relating to misuse of trade marks. They relate only to registered trade marks, not to unregistered trade marks or get-up, and only to marks for goods, not marks for services.

It is a criminal offence to apply a mark identical to (or likely to be mistaken for) a registered trade mark to goods, labels, packaging or advertising materials intended for use in relation to goods without the proprietor's permission. It is also a criminal offence to sell, hire, offer for sale or hire, or to distribute the goods or the materials.

The penalties are substantial. On conviction in a magistrates court an offender can receive a 6 month prison sentence or a fine up to £5000 or both. On conviction in a higher court, the penalties can be 10 years or an unlimited fine or both. Directors and other company officials may be personally liable if they have connived in or consented to the commission of the offence (see commentary in Section 2.10.2).

It is also a criminal offence to claim that a trade mark is registered when it is not, but the maximum fine for this is £50.

6.5 Avoid being sued

It should now be clear to the reader that the best time to minimise the risk of being sued for infringement of a registered trade mark, for passing-off or even misuse of someone else's company name, is when a trade mark is selected in the first place. This is the time to carry out searches of public records to locate similar marks in the same area of trade.

The register of trade marks is publicly available online at www.patent.gov.uk.tm. Marks in a particular class and for particular specifications of goods can be searched to see what is already registered. The search can be made for identical marks and those likely to cause difficulties because they might be confusingly similar. Marks registered through the European Community office, see Section 6.6.3, should also be searched.

Information on unregistered trade marks is less easily available, but the Internet and trade and telephone directories can be searched, and some thought given to the names used by other companies offering similar goods and services. The Register of Company Names can be searched for the same or similar words if a word mark is to be used, see www.companieshouse.gov.uk and go to WebCheck. If there is a possibility of major export trade, equivalent searches can be carried out in other countries; usually the USA will be first choice for engineering products.

All the searches are available through a trade mark agent, patent attorney or a commercial searching organisation, and the results can be available within a few days for the UK, and a little longer for overseas searches.

The public records can also be used to keep an eye on other activities that might affect a trade mark owner. For example, another company might apply to register a similar mark for the same or similar goods, and watching services are available which will inform the owner and give the opportunity of objecting to registration. These services are available on a worldwide basis also.

6.6 Trade marks in other countries

6.6.1 National systems

In some countries, usually current or former British Commonwealth countries, the law relating to trade marks is very similar to that in the UK. For example, using a mark without registering it builds up legal rights which can be applied to stop the use of similar marks.

In other countries, including many in the EU, only a registration gives any worthwhile rights. Trade mark laws are national laws, and a separate registration is required for each country except Belgium, the Netherlands and Luxembourg which have merged their trade mark systems. Furthermore, as with patents a registration is granted to the first person to file, so a British company planning to trade abroad must decide at an early date in which countries trade mark protection will be required.

To make filing abroad a little easier, there is an international convention; if a trade mark application is filed in the UK and an application is filed in another country within 6 months, then the UK date is effective in that country. To some extent this limits leap-frogging by unscrupulous competitors who otherwise might register marks identical to the UK application in the countries of interest, and force either a change of name with inconvenient and expensive reprinting of labels, user manuals etc., or even a purchase of the rights in those countries.

On the other hand, a mark can be used before it is registered without invalidating the registration, so the priority system is not as important as it is for patents. A few countries, notably the USA and Canada, insist that a mark must be used in that country before registration is permitted, but an application can be filed before use, if the plan is for use to follow quite quickly.

In many countries there is no examination to check for similar marks. If the correct fees are paid, the trade mark is registered. It is possible for identical marks to be registered for identical goods, and then the two trade mark owners must resort to the courts to sort out the position – a lengthy and expensive process.

In some places a trade mark can be registered for a service, but other national laws allow registration only for trade marks to be used on goods. Registering a mark abroad is usually a little more expensive than in the UK, say £1000 if no major objection is encountered, especially if communications to and from the trade mark office must be translated into and out of English. If a mark is not registered, its use by third parties can sometimes constitute unfair competition, and legal action to stop the misuse can be taken on that basis.

6.6.2 TRIPS and trade marks

The TRIPS Agreement (TRIPS – Trade-Related aspects of Intellectual Property Rights) places on member countries, which have signed it, the obligation to make trade mark registration available for all visually perceptible signs that are capable of distinguishing the goods or services of one undertaking from those of another. Such signs must include personal names, letters, numerals, figurative elements and combinations of colours.

Countries may allow distinctiveness to be acquired through use of a mark, and can decide whether to allow registration of sounds or smells. A registered mark must give the owner the exclusive right to prevent use by a third party in the course of trade of an identical or similar sign for identical or similar goods, the aim being to prevent confusion.

TRIPS gives extra protection to well-known marks. After registration a mark must be used; if not, it can be cancelled after 3 years, unless there are mitigating conditions. The minimum provision is for the initial registrations, and for renewals, to be for a minimum of 7 years. Registrations must be renewable indefinitely.

Readers are reminded that different countries will phase in the arrangements at different speeds.

6.6.3 International trade mark systems

The main international system likely to interest engineering companies is based on the objective of having harmonised laws in the EU. It is run by the Office for Harmonisation of the Internal Market (Trade Marks and Designs) and marks registered by it are called Community Trade Marks (CTMs). The office is in Alicante in Spain.

A CTM is valid in all member countries of the EU. A single application is filed for a CTM via the UK trade mark office with an application fee of €975 for up to three classes, and is checked to see that it is capable of graphical representation, and is not inherently incapable of distinguishing. The CTM office does not check for conflicting marks, but trade mark offices of member states which have trade mark search facilities carry out searches for identical and similar marks, and the results are sent to the applicant without comment.

If the mark is not rejected at this stage, it is advertised and any person can oppose the registration within 3 months. Usually an opposition is based on an existing mark registered in a national system. During the opposition, the CTM office takes an active view on whether the two marks are too similar or whether a CTM can be registered. If a prior right exists in one member country, then a CTM cannot be registered.

When all differences have been resolved, the mark is registered, after payment of a substantial fee, currently €1100. The registration can be renewed every 10 years. If the owner already has one or more national registrations, they are held in abeyance while the CTM is in force.

A second international system is the Madrid Agreement. This allows anyone with a trade mark registration in one member country to make a single application covering all, or a selected number, of other member countries, currently 77 in number. The single application must be made within 6 months of the initial trade mark application and only if the mark has been registered. The filing is made in the International Office of the World Intellectual Property Organisation (WIPO) in Geneva, and results in a national trade mark in each designated country. The UK trade mark office keeps a special register of such marks.

6.7 Domain names

6.7.1 Introduction

Trade marks are registered in a specified way, for specified goods or services, in a particular country. Domain names can be used and viewed worldwide and are registered on a first-come first-served basis. Domain names are not really intellectual property as such, but the power of registered and unregistered trade marks to control misuse of company names and marks, and the enormous importance of domain names in today's world of e-commerce, make a short section on the subject essential to this chapter.

6.7.2 Early problems

In the 1990s, there was an explosion of names registered by US Network Solutions, the administrator of the top level domain (TLD) with the suffix .com. Two sorts of problems quickly arose: (a) owners of trade marks identical to the domain name objected to use of the domain name and (b) unscrupulous persons registered variations of famous company names and offered to sell them for large sums of money – this is called cyber-squatting.

In the UK at this time, a business 'One in a Million' registered many domain names including the company names of British Telecom, Marks and Spencer, Virgin etc. and offered them to the companies. Trade mark law came to the rescue. The Court of Appeal held that One in a Million had infringed the registered marks of the companies and had also been guilty of passing-off.

At first, the suffix .com dominated, but then country code TLDs were introduced. In the UK, the company Nominet controls .co.uk for companies; .net.uk for Internet service providers and .org.uk for non-commercial organisations. Similar companies in, for example, France and Germany control the equivalents for .fr and .de. Some countries insist on a business in that country before accepting a registration, others do not. There are now over 200 domain name registration organisations worldwide. A registration usually lasts for 2 years, and if not renewed, can be allocated to another owner.

6.7.3 *Introduction of international controls*

It quickly became clear that, whatever the philosophy of the Internet, controls were needed. In 1999, the non-profit organisation Internet Corporation for Assigned Names and Numbers (ICANN) agreed on a set of guidelines to approve new domain name registrars, and accreditation for registration of names with the suffixes .com, .net and .org. Several country code TLDs participated voluntarily. A Uniform Dispute Resolution Procedure (UDRP) was also agreed.

The UDRP is most often run by WIPO (the World Intellectual Property Organisation based in Geneva), which has dealt with the majority of disputes to date. The procedure is more like arbitration than litigation, with a case being heard by one or three experts. Fees are a modest US$3000 for three experts, but the legal fees for advisors are at much higher levels. The procedure is quicker and cheaper than going to court and is international. Information is available at www.icann.org and www.arbiter.wipo.int.

There are three criteria for winning a transfer of a name; the domain name must be identical or confusingly similar to a registered or unregistered trade mark; the use complained of must be by someone having no legitimate interest in the name; and there must be bad faith, such as deliberately confusing use, an offer to sell the mark or a history of registering names based on third party trade marks.

Under the UDRP a domain name can be cancelled or transferred, but there is no award of damages, and no power to stop another name from being registered.

6.7.4 *Recent developments*

In the USA there is a specific law, the US Anticybersquatting Protection Act, which is the basis for legal action against the registration of a domain name identical or similar to a US registered trade mark, and which gives some protection to personal names. Action can be successful even when the actual owner is well hidden by Internet tricks and cannot be traced.

At the time of writing this book, there is no plan for a similar law in the UK.

In addition to .com, .org and .net, there are the suffixes .pro for professionals – for example, .pro.med for doctors – and .eu for Europe-based businesses. There is also .tm for trade marks, with the unusual and practical feature of a registration under it having a life of 10 years, in line with trade mark registrations, and not the usual 2 years.

6.7.5 The future

Since the initial free-for-all, most businesses of any size have introduced a domain name policy and have registered their own names. The law continues to develop, for example, in the area of metatags, when a link is made to another website without showing the Home Page, causing confusion as to whose site is on view.

With over 200 country code TLDs, and with an expanding number of suffixes, and the many variations of names when addition of a hyphen or underscore counts as a different name, it is just not possible for even the largest company to register all possible variations of its name and trade marks as domain names. The best option is to register trade marks in countries where unauthorised activity is likely to be most damaging.

6.8 Summary of trade marks

The aim of a trade mark is to generate repeat business by the use of a memorable name or symbol. Trade marks can be used for goods or for services and are preferably registered at the Trade Mark Office although use of an unregistered trade mark builds up valuable goodwill.

An effective trade mark is one which does not describe in any way the goods or services it is to be used for. Preferred examples are 'made-up' words and words taken completely out of context.

A trade mark can be registered if it is not misleading and not too similar to existing marks. The registration can be renewed indefinitely by payment fees at intervals of 10 years. Such a registration is a useful weapon against misuse of domain names by others.

Chapter 7

Ownership of intellectual property rights and rights of employees

The engineer who creates any one of the six types of IPRs (intellectual property rights) described in previous chapters is not necessarily the owner, especially if that engineer is an employee. In addition, there are rules of ownership when a right is created for payment, which are different for different rights. The general position on ownership is set out here, and the mechanisms for changes of ownership are explained.

If an employed engineer makes an especially successful invention there is a right to a reward from the employer in certain circumstances. This is reviewed in the last part of the chapter.

The chapter applies only to engineers employed in the UK, or working abroad but attached in some way to a place of business in the UK. For normal overseas employment, local law will apply, which is outside the scope of this book.

7.1 Ownership

7.1.1 Necessary definitions

Three general terms are widely used and need a few words of explanation.

7.1.1.1 'Employee'

In most cases it is clear whether an engineer is employed. An employer has control over what an employee does and when and where the employee works. Employees receive holiday pay, their employer pays national insurance and sickness benefits, and sometimes pension fund contributions.

If there is any doubt, then the arguments will almost certainly be parallel to those used in relation to income tax, when someone tries to reduce payment by claiming to be self-employed; in such cases the tax authorities look at all the circumstances. Calling oneself a 'consultant' will not sway the authorities if it is not true. A consultant is in business on his or her own account. An employer and an employee have mutual obligations.

Legally, every employee is bound by a 'contract of service' (usually a written contract these days, but the contract can be verbal or implied); a self-employed person may work under a contract for services, but again titles do not matter, facts do.

Lecturers in higher education, who are mainly employed to teach but who may also do research, may raise some questions as to their employment duties. Guidelines produced by the Committee of Vice Chancellors and Principals recommend that all lecturers should have the duty to carry out research. Individual establishments may have their own rules, often more generous to an employee than statute law is.

Engineering apprentices may or may not be employees, there is no related legal decision to help.

Holding shares in a company is irrelevant: a company is a separate entity from its shareholders or members, so one can hold shares in a company and still be an employee of that company.

Company directors have wide obligations, as do partners in a partnership. As far as ownership of IPRs is concerned, directors and partners usually hold the IPRs on trust for the company or the partnership.

7.1.1.2 'Commissioned'

Some of the laws relating to ownership of IPRs refer to 'work commissioned for money or money's worth'. A commission is simply an order or authority to do an act, not necessarily in writing. Actual payment can be made, or the work can be carried out in return for other services or for the supply of goods, which would constitute 'money's worth'.

7.1.1.3 'Person making the necessary arrangements'

For several different types of IPR, the right is owned by a 'person making the necessary arrangements' for the work to be created. For engineers, the most important use of the phrase occurs with computer-generated works, in which case the words only apply since the Copyright, Designs and Patents Act 1988 came into force on 1 August 1989.

Help on interpretation is given from use of the phrase under the previous copyright act with respect to copyright in films, where it is taken to mean the film producer, and often indicates a financial authority of some sort. In the computer area, the phrase probably means the company or the person owning the computer on which the computer-generated work is created.

7.1.2 Ownership of patents

As far as patents are concerned there are three separate persons, the inventor, the applicant and the owner of the patent rights. In the legal sense, 'a person' can be a human being or a company.

The inventor is the person who has the creative idea that constitutes the invention. Often there are two or more inventors, each having genuinely contributed to the invention. Merely giving advice or carrying out tests or managing the inventor's section or department does not count as joint inventorship.

The applicant is the person who files the patent application, which need not be either the inventor or the owner. Anyone can file an application, but a patent can only be granted to the inventor or the correct owner. Any disputes regarding ownership must therefore be sorted out while the patent application is proceeding through the patent office.

Ownership is usually judged in an employment context; the position is that patent rights frequently belong to the employer. Under the Patents Act 1977 there is a three-part test. One first defines the 'normal duties of the employee', and then decides if an invention was likely to result from these duties. One also looks at any special obligations of the employee at the time the invention was made.

7.1.2.1 Normal duties

'Normal duties' may be defined in the engineer's contract of employment or in pre-employment correspondence. However, duties often change with time and are not necessarily reflected in the records, so one also needs to consider the employee's general 'duty of good faith' to the employer, that is, the obligation to further the employer's business, and also the custom and practice in the company. One must also consider if the engineer was working on a special project in an unusual area at the time the invention was made. From all these facts, a judgement is made about what the engineer was really expected to do at the time the invention was made.

7.1.2.2 Is an invention likely to result?

Once the engineer's duties have been defined, one decides if this is the sort of work from which an invention is likely to result. Much engineering work is creative and therefore it is highly likely to generate inventions and any such invention will belong to the employer. Other types of work where inventions are not expected would, for example, include production of engineering drawings, where a draughtsperson is employed to turn sketches into detailed drawings; if such a person makes an invention relating to a product in one of those drawings, it is highly probable that the employer would not be able to claim ownership of that invention.

7.1.2.3 Does the employee have special obligations?

This phrase is taken to refer to senior management, who have very broad responsibilities to their employer, and whose inventions in any field relating to the company would probably belong to that company. How far down the hierarchy of the company this provision extends is not clear. A managing director's obligations would extend over the whole of the company's activities, but those of a sales manager would not.

The three-parts test was considered by the High Court in 1984 in a case relating to Wey valves. The inventor, Mr Harris, was employed by Reiss Engineering as manager of their Wey valve department. The company sold valves made by a Swiss company, Sistag, or valves made by Reiss using Sistag's drawings. Reiss did not have any development facilities, did not design valves, or make improvements or modifications to them. While Reiss advised customers on which valve body material and sealant to use, any problems developing after sale of a valve were referred to Sistag.

Mr Harris developed a new kind of valve, which solved a problem encountered when Wey valves, and other types, were used for powders, such as pulverised fuel. He applied for a patent, and Reiss claimed ownership.

The court found that the invention belonged to Mr Harris. His normal duties did not include designing or inventing, and indeed the company itself took no responsibility for designing or inventing Wey valves. Considering his status, Mr Harris did not have 'special obligations' towards the company; his obligation was limited to selling Wey valves. While a managing director would have an obligation covering the whole spectrum of a company's business, a sales manager's obligations are much more limited.

The way the law is expressed puts the onus on the employer to prove what the employee's duties were, and to show that an invention was likely to result. But the invention does not have to be made during working hours or on the company's premises: an engineer's brain is owned by an employer 24 hours a day, so having an idea at the weekend or on holiday does not affect the answers to the questions set out above. Even testing the idea at home first, using one's own materials, does not alter the ownership position. Similarly, if an engineer tests a personally owned invention using an employer's equipment and materials, this does not allow the employer to claim ownership, although there might be differing views about misuse of company property and time.

The law on ownership of patentable inventions cannot be varied by an agreement signed before the invention is made. Therefore, an employer cannot insist that an employee signs away any rights in advance, for example, when arriving at a new job. Agreements can be made after the invention is created which determine who owns it, the employer or the employee, presumably because the inventor then has some idea as to how important the invention is, and can therefore make a reasoned judgement on its value.

Before the Patents Act 1977 came into force employers could be quite grasping, and one case is a useful example. A storeman, Mr Hudson, was employed by Electrolux Limited, which among other things made vacuum cleaners. Mr Hudson and his wife jointly invented a device for holding the disposable paper bags in a vacuum cleaner which allowed any type or shape of bag to be used. Mr Hudson had signed an employment contract agreeing that all inventions made by him would belong to Electrolux. The High Court found that this was unacceptably broad; he was not employed to make inventions and the rights belonged to him, not his employer. The current law makes it clear that such a contract nowadays is not enforceable.

The previous law was weighted much more towards the employer as far as ownership of patents was concerned. It was based not on statute law but on the general law of 'master and servant', under which anything produced in the course of employment belonged to the employer. The words came to mean anything authorised by an employer, whatever the actual duties of the employee. Only the extremely unfair type of contract signed by Mr Hudson was held to be unreasonable, so the Patents Act 1977 introduced a more balanced position.

In addition to joint inventorship rights there can also be joint ownership, for example, by two companies. There can be joint ownership between an employee and an employer by agreement after the invention is made, but there cannot automatically be joint ownership by an employee and an employer. The disparity of influence would be too great, and the law does not award this position.

7.1.3 Ownership of copyright

This section applies to copyright created since the Copyright, Designs and Patents Act 1988 came into force. Copyright is a long lasting right; it is for the life of the author plus 70 years but before 1 January 1996 it was life plus 50 years; so previous copyright acts might still be applicable. The position on ownership has not changed substantially, but a precise date of creation may need to be determined to apply subtly different legal wording.

Because copyright protection is automatic with no mechanism for registering rights, there are only two categories of legal people, the author and the copyright owner.

The word 'author' is used universally as far as copyright material is concerned and includes designers, artists and composers. It means the actual creator of the work, which depends on the sort of work that is being protected.

For written copyright, the author is the person putting pen to paper or using a PC. For copyright in drawings, the author is the person who fixes the picture on whatever medium is used – the person drawing the sketch, the draughtsperson preparing the detailed drawing or the engineer using the CAD program. If a computer program is being written, the author is the person using the keyboard. For a copyright work generated by computer without human intervention the 'person making the necessary arrangements' defined in Section 7.1.1.3 owns the copyright.

In fields more distant from engineering, the photographer is the author of a photograph and the architect is the author of a drawing of a building, but if a builder builds without making a drawing, the builder is the author. For a film, the producer is considered to create the copyright, not the camera person; for a sound recording, the person creating the work is the author; for a broadcast, the person who has responsibility for its content is the author.

If the author is not an employee, the author automatically owns the copyright. If the author is employed, then the test for ownership of copyright is much less complex than for patents. The law simply states that copyright material generated 'in the course of employment' belongs to the employer. The questions to ask would be:

- Is the engineer an employee?
- Was the copyright work created as part of that employment?

This judgement is sometimes difficult to apply when an engineer's work and outside interests overlap. This is frequently the case with software engineers, who often write programs for their employers and for other contacts, using the same skills.

If the applications of the programs are quite different, such as process control software for an employer and a computer game for a third party, there will be little risk of conflict and the engineer will personally own copyright in the game.

But if the applications do overlap disputes arise. This was the case with Mr Magee, who at one time was employed by Missing Link Software as the manager of a team writing software for a personnel management system. A few months after Mr Magee was made redundant by Missing Link, a competitor started to market a very similar system, running on the same hardware; it had been written by Mr Magee.

Missing Link claimed ownership of the copyright. An independent expert formed the opinion that given the size of the program and the timescale, the similar system must have been written while Mr Magee worked for Missing Link. Missing Link argued that he had been employed to work in the field of personnel management software systems, so any such software that he wrote was part of his employment duties and belonged to his employer. The High Court judge held that this was a strong argument and granted an injunction to stop the competitor from continuing to market the system. (NB: There was no argument of copyright infringement, the programs were quite different creations.)

If the author of a copyright work is not employed but has special obligations to the company, for example, as a director, copyright is considered to be held in trust for the company and the company could request formal ownership if it wished. Similarly, the copyright created by a partner would be held in trust for a partnership. An employer and an employed author can agree in advance that the normal rules of ownership do not apply, and that the employee owns copyright.

Although copyright in commissioned material does not automatically belong to the commissioning company, if one company pays another to create copyright material, the circumstances may be such that the creating company holds the copyright in trust for the commissioning company. That company, having paid for the work to be generated, will at the very least have the implied right to use the work for the intended purpose, for example, by copying it for internal use within the company or for sale, depending on the intention when placing the contract.

The implied right may not be unlimited. For example, if a photograph is commissioned for limited use, such as an internal record, and the company then decides to use it for a major advertising campaign, the photographer might be able to argue that the implied right is limited to internal use; for an advertising photograph, the fee charged might have been higher.

7.1.4 Ownership of rights in registered designs

This section applies to registrable designs created since the Registered Designs Act 1949 came into force, as amended by the Copyright, Designs and Patents Act 1988.

The person who creates a registered design by putting pencil to paper or operating a keyboard to run a CAD program is the author of the design. The owner of a computer on which a computer-generated registrable design is created would be the person 'making the necessary arrangements', so the author can be a company.

The relevant law assumes that a registrable design is created by one person only, so if there are two or more creators, they need an agreement about how to deal with joint ownership.

The author is also the owner of the rights in a registrable design except under the following circumstances:

1. The design was produced 'in the course of employment' – the words are the same as those used for copyright. There have been few cases where any doubt is raised as to whether the work was done in employment or not: most registered designs belong to the employer.
2. The registrable design was produced under a commission, when the person placing the commission is the first owner. This means that if Company A pays Company B to carry out design work by a company employee, Company A owns the registrable design from the moment it is created.

The law does not make any reference to written agreements varying the ownership position either between employee and employer or in the context of a commission, but agreement in advance is not forbidden and is therefore presumably possible.

7.1.5 Ownership of design rights and topographies

This section applies to designs created since 1 August 1989.

The designer of a design right is the engineer that creates it and if two or more engineers create it they are joint designers. The owner of the computer on which a computer-generated design right is created is also 'the designer', so a designer can be a company.

The designer is also the owner of the design right unless the design was produced by an employee or under a commission, when the employer or the person or company placing the commission own the rights.

The statute law does not forbid an employee and an employer agreeing a different ownership position, so this is presumably possible. The position under a commission can also be varied by written agreement.

Ownership of rights in topographies follows the same rules as for design right, except that the statute law specifically says that an employer and employee or a commissioner *can* vary the ownership position by a written agreement.

7.1.6 Ownership of trade marks

There do not seem to have been arguments between employees and employers about ownership of trade marks, which is a little surprising given the high values that are associated with well-known marks. If an employee creates a logo the rules for ownership of artistic copyright would apply. Single words and short phrases do not attract literary copyright. Often trade marks are created by agencies in return for payment, and the contract terms generally assign all rights to the company intending to use the mark. The owner of a trade mark is generally the person or company that uses the mark first, or is the first to apply to register it.

Where disputes do arise is in conflicts between similar marks where either one is used without registration and the other is registered later, or both are used, possibly in a small way, and then one business expands until the risk of confusion becomes noticeable. This can be sorted out by agreement between the owners, for example, limiting use to a specific geographical area for a small business, or to different goods or services. If one or both of the marks are registered, these limitations can form a condition of the registrations.

7.1.7 Ownership of confidential information

For the more complex judgements to be applied in testing ownership of confidential information see Chapter 5.

7.1.8 Problems of joint ownership and split ownership

7.1.8.1 Joint ownership

If any type of IPR is jointly owned the joint owners can each use the right themselves. Thus, each joint owner can make a patented product and each can copy copyright material. What they cannot do is sell the right or grant licences to third parties without the permission of the other joint owner(s).

This is the reason that joint ownership by an employee and employer is not automatic in any intellectual property (IP) law, although it seems at first to be the logical way to solve ownership disputes. In such a case, the employer could use the patent or copyright material within the company, but the employee could use the right only by licensing it, which would need the permission of the employer. This is clearly unequal.

The restriction also needs to be considered when non-manufacturing institutions, such as higher education establishments or research organisations, are working in collaboration with manufacturing companies. If IPRs are jointly owned, the manufacturing company can use the rights internally without the permission of the joint owner, but the non-manufacturer can only use them through a licence with consent of the manufacturer.

The simple solution is to give ownership to one party and explicit licence rights to the other party. For example, the manufacturing company can own the IPRs, with a university having the right to licence third parties. The licence may need to be limited to a field of use which avoids direct competition with the manufacturing company that has funded the work, but this is usually achievable.

7.1.8.2 Split ownership

This is when different types of rights are owned by different parties, for example, an employer and an employee. Suppose an engineer invents a new product which is patented and the rights belong to the engineer individually; the engineer will almost certainly have prepared drawings of the product, and since the tests for ownership of copyright and design right are less specific than for patentable inventions, the employer may well own those rights.

In these circumstances, the law is explicit. The employee can use the invention and the copyright and design right also. The employer has no rights to stop such use as far as exploitation of the invention is concerned. What the employee cannot do is to use copyright or design right created by fellow employees as part of their employment, for example, if detailed product designs were produced by someone else. The employer has full rights over those designs. Also, the employee cannot disclose the employer's confidential information, either in filing a patent application, or in exploiting the invention.

7.2 Changes of ownership

Once the first owner has been established, any type of IPR can be sold (assigned) into new ownership provided certain formalities are completed, or a personally owned right can be bequeathed in a Will. After any IPR has been assigned, the previous owner has lost all control of it and cannot use the invention or copy the copyright material etc. or license others to do so.

The position of joint owners is that although each can use the IPR without the permission of the other, the share in the right cannot be assigned (or licensed) without the permission of the other joint owner(s).

Intellectual property rights can also be mortgaged (see Section 7.2.2).

7.2.1 Assignment of individual IPRs

7.2.1.1 Patents

Rights in a patent or a patent application can be assigned, provided the sale is recorded in writing and both parties to the deal sign the document. The assignment can be registered in the Register of Patents. Such a registration is not essential, but it is highly advisable because if a patent is first assigned to Company A who does not register the change of ownership, and then the original owner (dishonestly or in error) assigns it to Company B, Company B would not be affected by the earlier unregistered assignment. Furthermore, if the assignment is not registered within 6 months of the sale and if the new owner sues for infringement, there is no right to receive damages for any infringement committed while the assignment was not registered.

The right to file a patent application can also be assigned. Any person has the right to file an application so a formal record of the assignment is not essential, but it is helpful if there is a subsequent dispute. In an employment context, a written confirmation that an invention was made as part of an engineer's normal duties can be a useful record as it helps to avoid later disagreements. If a US patent application is filed, a formal assignment of rights from the individual inventor is always essential because only the inventor can apply for a US patent.

7.2.1.2 Registered designs

A design registration, an application for a registration or the right to make an application can be assigned. The assignment must be registered in the Register of Designs;

if not, the assignment is not admissible in court as evidence of ownership. However, there are no restrictions on payment of damages as with patents.

Since registered designs and design rights are so closely linked, to eliminate any doubt, it is advisable to include the design right associated with the registration in the assignment.

7.2.1.3 Design right

A design right can be assigned provided there is a written agreement signed by the assignor, that is, the original owner. The person receiving the rights need not sign. The right can be assigned in advance, that is, before the design is created, provided the person who would otherwise be the first owner signs the document.

If a design right is assigned and the same owner owns a registration for the same design, the assignment is taken also to mean assignment of the registration unless the document makes it clear that this is not the intention.

7.2.1.4 Copyright

Copyright can be assigned provided there is a written document signed by the assignor, the person giving up the ownership. The assignment can relate to future copyright, that is, material not yet generated, provided the person who would otherwise be the first owner signs the document.

If the copyright is artistic copyright in drawings of an article in which there will be design right and may also be a design registration, and if the work is commissioned so that design right and any registration belong to the person paying for the work, then surprisingly copyright in the drawing does not automatically change hands. It is clearly convenient if copyright is formally assigned into the ownership of the commissioning organisation.

7.2.1.5 Moral rights

Moral rights can change hands only as a bequest in a Will.

7.2.1.6 Trade marks

Registered trade marks are usually closely associated with 'goodwill', that is, the general reputation of a business and it is usual to assign registered trade marks with the goodwill. If an assignment document refers to transfer of goodwill, the associated trade marks are transferred by implication.

Trade marks can be assigned without goodwill provided the result is not deceptive or confusing, that is, the result is not that two companies are using the same or similar marks relating to the same or similar goods or services.

7.2.2 Mortgages

Intellectual property rights can be mortgaged and in recent years it has become increasingly common to use IPRs as security for a loan. The IPR can either be the subject of a mortgage in the normal sense in which legal title passes to the mortgagee, or alternatively a charge can be created over it when there is no passage of legal title, but certain rights are given as security for the loan.

Mortgages for patents, copyrights, registered designs and design rights cause few problems. In effect the right is assigned. Mortgages for trade marks are technically more difficult (because the owner of a mark is supposed either to use the mark himself or herself, or to control use by others under licence), but are not impossible.

7.2.3 Transfer formalities

When any IPR is assigned, it will almost always be in return for payment of some kind, in legal terms 'a consideration' (although this can be as little as £1), so the assignment constitutes a contract. A contract is valid if it is signed by an individual engineer acting on his or her own behalf. Sometimes the signature is witnessed; this is merely to provide additional proof if the contract is ever disputed.

Since 1989 there has been no need for a company to have a common seal for use on documents: a contract signed by a director and a company secretary, or by two directors, in each case signing on behalf of the company is legally valid. This applies unless the company itself has internal regulations which vary the general position.

In the rare cases when an IPR is assigned without consideration, the assignment must be by way of a deed. The appropriate wording must be used for individuals and for companies, and the signatures must be witnessed. The professional legal adviser involved in the transaction will take care of this point.

After a transfer of a patent, registered design or registered trade mark, the assignment should be recorded at the Patents, Designs or Trade Marks Registry as appropriate.

7.2.4 When do IPRs change hands?

The preceding paragraphs cover the sale of a single type of IPR but frequently a bundle of IPRs changes hands. The situations when such an assignment occurs are reviewed in Chapter 10.

7.3 Employee's compensation under the Patents Act 1977

7.3.1 When is compensation available?

While one hears of 'master patents' and of companies making huge sums of money from an invention, these are very rare cases. Most patents do not cover their costs, much less make a profit, but if vast profits do arise, then the employed inventor who made the invention has the legal right to a share. This applies to both UK and foreign patents.

Two sets of circumstances are covered by the Patents Act. Either the invention and the patent were always owned by the employer, using the tests described in Section 7.1.2, or the employee originally owned the patent and either assigned it, or exclusively licensed it, to his or her employer (NB: An exclusive licence excludes the owner from using the invention).

If the employer owned the rights and the patent was of outstanding benefit to the company, or if the benefit to the employee from the assignment or licence to his or her employer can be regarded as inadequate in relation to the benefit to that employer, then it may be justifiable for the employer to pay compensation to the employee.

7.3.2 Some relevant tests

7.3.2.1 'Outstanding benefit to the employer'

The benefit of a patent to a company can take several forms. The most usual is that a patent allows the sale of a patented product or the use of a patented process and keeps competitors out of the market. Another possibility is that a patented invention is not actually used by a company, but the existence of the patent blocks a competitor, so the patent owner benefits indirectly. Yet another alternative is that the company licenses the invention or assigns it, in either case in return for payment.

The benefit to the company must be 'outstanding'. Interpretation of the word is not easy. The law states that the size and nature of the business must be considered, so a very large company would need a much higher benefit than a medium sized company before the word would be applicable. This was demonstrated in 1991 in a case involving British Steel.

Mr Monks, an employee of British Steel, invented a rotary valve for use in steel making; it controlled metal flow from a melt vessel. It had been described as 'a miracle valve' and had won a Queen's Award for Technical Achievement. British Steel had filed patent applications in many countries but had not received any licence income and for various reasons had used the valve only at its facilities on Teeside. British Steel had paid the inventor £10 000, but he applied for additional compensation.

The benefit to British Steel by reduced processing costs, lower scrap losses and increased output of the Teeside plant was accepted by the patent office to be £100 000–200 000 a year. But this was only 0.01 per cent of British Steel's annual turnover and the decision was that this was not 'outstanding'. Mr Monks had argued that the benefit was much higher, about £6 million per year, but even if this figure had been accepted as justifiable, it would still not have been an 'outstanding benefit' to a company the size of British Steel.

In a later case, Memco Med, it was said that the implication of a superlative in the British Steel case was perhaps too high, although the benefit had to be greater than merely substantial. In a decision by the Comptroller of Patents, the possibility of considering future benefit was not ruled out.

7.3.2.2 Benefit from the patent

It must be the patent, not just the invention, that is the source of the benefit.

In a case involving a wide-angle head-up display system for a fighter aircraft, sales of the patented system were very good and the inventor claimed for compensation. The claim was brought under a US patent for sales in the USA and this was held to be perfectly acceptable. The Company, GEC Avionics (formerly Elliott Brothers Limited), showed that an earlier version of the system falling outside the patent had a similar level of sales. The argument was that the patent did not form the basis for success, so compensation was not justifiable.

Another point to consider is that if a company is dominant in the market place for other reasons, the dominance might be the real basis for success of the invention with the patent protection contributing very little. It is also possible that the success is derived from a particularly good design, or even an exceptionally successful advertising campaign, rather than the patent.

The patent need not be shown to be valid. If a company benefits from protected sales or from licence income for some time and the patent is later found to be invalid, that company has still benefited from the earlier protection.

An employing company cannot escape its responsibilities for paying compensation by assigning a patent and retaining the protection, but correctly claiming that it is no longer the owner. For example, if a patent is assigned to an associated business in a large group of companies, or in the case of a small company to a relative of the owner, then the wording of the act is such that the employer is still obliged to pay compensation.

The effect of requiring that the benefit derives from the patent and not the invention is that if an employer decides not to file a patent application, for example, because the invention is of a type which can be kept secret, such as an improvement to a process, then the employee does not have any right to compensation.

7.3.3 Amount of compensation

Whether the invention was originally owned by the employer or by the employee, the requirement is that the employee has a fair share of the benefits. All the surrounding circumstances must be considered, with a number of factors spelled out in the Patents Act.

When the employee owns the rights, one needs to take the following factors into consideration:

1. The employee's duties in the employment (the higher the job level, the lower the compensation); the employee's salary level (the lower the salary, the higher a payment may be); any other advantages the employee has received (e.g. a promotion).
2. The effort and skill the employee devoted to making the invention – persistence in an unpromising situation should be rewarded.
3. The effort and skill contributed by other people; some contributions are not sufficient to qualify as joint inventorship but can still be an essential part of success; this type of judgement can cause difficulties if success is a team effort, but if the inventor has received a considerable amount of assistance from colleagues, the compensation would be reduced.
4. The contribution of the employer – were expensive facilities used or were the employer's managerial and commercial skills called upon to make the patented invention a success? If so, compensation will be set at a lower level.

If an employee-owner has licensed or assigned a patent to an employer, similar factors outlined below are to be considered.

1. The conditions in the exclusive licence.
2. Whether the invention was made jointly with another employee, for example, an employee whose duties were such that the employer owns the rights. If this is the case, then the employee claiming would receive a lower compensation.
3. The contribution of the employer: were the company's facilities and managerial and commercial skills used in exploitation of the invention? If so, payment will be reduced.

For both employer-owned and employee-owned inventions, all the other surrounding circumstances will be considered when compensation is assessed. The compensation can be money, or money's worth, such as the provision of a car. Compensation is assessed on the benefit the employer has received to date and any future benefit which is reasonably predictable. If the prediction turns out to be an underestimate, a second and subsequent applications for additional compensation can be made. Monetary compensation can be paid as a lump sum or as a recurring sum; the aim should be to reduce the employee's tax liability.

Compensation is only paid when it is just to reward the employee; this would not be the case if, for example, the employee had misused the employer's confidential information in applying for a patent.

7.3.4 How to claim

Almost all claims of this nature are settled internally. Large, well-regulated companies often have a Patents Award Scheme to deal with employee compensation. If there is no amicable settlement then the employee can apply to the patent office, see www.patent.gov.uk or the Patents County Court or the High Court, see www.courtservice.gov.uk. Legal aid will be available for the second two courts (search 'legal aid' on the website) although the levels of income and savings at which legal aid applies are very low indeed, and few inventors are likely to qualify even if they are no longer in employment. Usually such formal claims are made by ex-employees because current employees are rarely willing to take the risk of upsetting their employer.

If there is a formal application of this nature, it must be made after a patent is granted but within 1 year of its lapse, whether this is by failure to pay an annual renewal fee or by the patent reaching the end of its 20 year life. So if the employer stops paying renewal fees, the employee needs to know. This information is available by filing a form at the patent office asking to be warned when the situation arises and paying a fee; this is known as a caveat.

If formal proceedings are begun, the employee-engineer needs detailed information to support the claim. This can be achieved by the normal process of legal disclosure of documents through formal requests to the court. In this way an employee can gain access to financial records, but disclosure is a slow process and it may take years to be allowed to see all relevant records.

Since patents last for 20 years, a claim can be made literally two decades after an invention was made. The importance of good records is clear. Every engineering company needs to record who worked on an invention in addition to the inventor, and the level of assistance given by the company. The compensation provisions under UK Patent Law extend also to overseas patents, when similar timescales apply.

An employing company cannot force an engineer to sign away in advance his or her rights to compensation for outstanding patents. If a document which purports to do this is signed, that particular part of the contract is legally unenforceable. An employed engineer can sign away rights after an invention is made; presumably the value of the invention can be judged to some extent at this stage, although the likelihood of the benefit being outstanding will still be extremely hard to estimate.

The provisions for compensation to an employed engineer do not apply if there is a 'relevant collective agreement', that is, an agreement between a trade union to which the engineer belongs and the company, which relates to compensation for patented inventions: such agreements are very rare.

Close interpretation of the wording of the Patents Act 1977 gives some scope for arguing that the right to compensation extends to other types of IPRs as well as patented inventions, but the theory has not been tested in court.

7.4 Suggestion schemes

Large companies often have a formal Patents Award Scheme to deal with *ex gratia* payments for patented inventions, preferably at a high enough level to avoid the embarrassment of employees applying to the court for further payments. Companies may also have a suggestion scheme to encourage employees to put forward all types of ideas. Such a scheme can be very beneficial in improving not only efficiency but employee morale, provided the scheme has well-designed rules, is properly administered and is extensively advertised to staff. It is important to keep employees fully informed about the progress of an idea through the consideration stages, to treat all suggestions seriously, and to give rapid replies.

The preceding sections of this chapter will have made it clear that a company cannot assume that all the ideas its engineers produce are company property, especially if they constitute patentable inventions. Companies therefore need to proceed on a proper basis. All suggestions put forward by employees should be kept confidential, so that an employing company does not jeopardise the chance of an employee applying for a patent, if it turns out that the invention is the employee's personal property.

The majority of suggestions, even in an engineering company, will not be patentable, and will relate to minor changes to equipment or internal procedures when IPRs will be largely irrelevant. The ideas will still often be of benefit to the company, and some employers make a nominal payment for any suggestion, followed by a substantial payment for those put into practice.

The most important way to encourage innovation, at whatever level within a company, is management attitude. Japanese-owned companies tend to have a very positive approach to employees' suggestions, and therefore a very high-flow rate of ideas, compared to more traditional British companies.

7.5 Summary of ownership

If any material protected by copyright, design right, registered design or topography right is created by an employee as part of his or her normal work, the employer owns the legal rights. For patentable inventions, the employer owns rights in an invention only if the employee's job is of the type in which inventions are likely to be made. All other inventions belong to the employee.

If a design right or registered design is created under a contract, the company paying for the work automatically owns the rights, but not the associated copyright. Paying for the work does not automatically change the ownership of copyright or patents, they must be specifically assigned in a written document.

Chapter 8
Unfair competition

Revised and updated by Matthew Gream

8.1 Background

Intellectual property rights (IPRs) grant exclusive rights with limited exceptions. This is generally intended to incentivise and reward the investment in creating such properties, yet not restrict the operation of society nor hamper its economic progress. For patents, the exclusive rights are granted in return for public disclosure of the invention. For copyrights, there are fair dealing allowances for private and non-commercial study. For design rights, no protection exists in the parts that interconnect with other parts. Examples exist for other IPRs.

In maintaining this balance, economic competition is assured. However, there remain avenues for abuse that can distort competition. For example, a patent owner might hold a bundle of essential patents in a field of technology, yet refuse to grant licences on reasonable terms, thus stifling innovation. Elsewhere, a trade mark owning company might use its power as a well-known and successful franchisor to include non-essential conditions in a licensing agreement that allow it to control the supply of goods. In these cases, legitimate competitors are unfairly disadvantaged, and the incumbent monopolist is able to maintain its position with greater ease.

The early international treaties on intellectual and industrial property rights recognised the danger of unfair competition and required that countries address the problem. The Paris Convention of 1883 defines 'unfair competition' as 'any act of competition contrary to honest practices in industrial or commercial matters'. The TRIPS Agreement (TRIPS – Trade-Related aspects of Intellectual Property Rights) of 1994 requires the control of 'anti-competitive practices in contractual licences' and allows use of 'appropriate measures ... to prevent the abuse of intellectual property rights ... or ... practices which unreasonably restrain trade'. The proper functioning of IPRs is ineffective without these types of controls.

Historically, anti-competitive behaviour has been addressed both by the IPRs themselves – such as the UK Patents Act which allows the comptroller to grant a licence if the owner refuses to grant one on reasonable terms – or by specific laws. The UK has for a long time maintained laws against restrictive trade practices, not to mention the common law of equity, but these have evolved without particular intent, and have not been pursued vigorously as policy.

In recent years, however, a large body of law and practice has developed to address anti-competitive behaviour in all of its forms. These 'rules on competition', as they are called, apply to IPRs, and it is now, more than ever, necessary to continually balance the exercise of IPRs with competition issues.

The EU has not only played a leading and proactive role in developing and enforcing the rules on competition, but the nature of the EU with its differing Member States has resulted in a related area of anti-competitive behaviour concerned with the free movement of goods. It is a fundamental principle of the EU that there are no barriers to trade between states, and that national laws – such as those that apply to IPRs – do not hinder such trade. This principle is present in an ever-increasing body of EU regulation and case law.

Many other advanced countries have similar concerns, such as the USA with its anti-trust activities. These authorities have begun to work together to ensure that there is no escape from the rules on competition, whether on a national, regional or international stage. High-profile anti-trust cases that have been common in the USA, such as the break-up of the telephone company AT&T, are now a regular occurrence in the EU. Recently, the EU fined Microsoft a record €497 million for abusing its dominance in the market for operating systems.

It is simply not possible to understand or utilise IPRs without taking matters of competition into account. Failure to do so could be very costly. At the same time, the rules can be very complex and detailed, so the use of a specialist is always required.

8.2 The rules on competition

8.2.1 Background

The EU and UK rules on competition were once different in principle and practice, but after a period of intentional harmonisation, they are now largely integrated and should be considered as one.

The EU rules on competition stem from the EC Treaty which establishes the European Community as an economic area for the free movement of goods. The rules are designed to prevent anti-competitive behaviour that would restrict this freedom and so they are widely applicable. It has been found that anti-competitive behaviour between enterprises can have an effect on trade, even if that behaviour is confined to one Member State.

Articles 81 and 82 of the EU Treaty specify the competition rules that are the most relevant to IPRs. The rules are administered by the EU Competition Directorate (DGCOMP) and are primarily implemented by the recent EU Regulations 1/2003 and 139/2004. A large number of more detailed regulations and notices apply to specific circumstances, such as licensing agreements, R&D agreements, vertical agreements and subcontracting agreements along with clarifications on key concepts and approaches.

The UK rules on competition are embodied in the Fair Trading Act 1973, the Competition Act 1998 and the Enterprise Act 2003. Various aspects are administered by the Department of Trade and Industry (DTI), involving the Office of Fair Trading (OFT) and the Competition Commission (CC). The shape of these laws and their practice reflects harmonisation with the EU.

In the UK, new consumer oriented anti-competitive powers have been given to the OFT, while the EU drives an active programme to promote consumer awareness of competition issues – there is even a regular Competition Day for European citizens. These, along with a number of very public and high-profile cases, have ensured that consideration for anti-competitive behaviour is an increasing feature in all commercial, engineering and consumer activities.

8.2.2 Application

The rules on competition are largely divided into two distinct areas: those addressing agreements, such as a patent licensing agreement, and those concerned with dominance, such as an enterprise that has a monopoly as a result of owning key patents in a field of technology. The rules have a very wide scope as they apply to all forms of transactions and behaviours of any legal entity, whether a person or an enterprise.

In the early days of the EU, registered rights were regarded with strong disfavour. Subsequently, it was realised that innovation was discouraged without the availability of effective IPRs. The attitude to IPRs has been more commercially oriented in recent years and is still evolving. The relatively recent enactment of the Community Trade Mark and the two Community Designs have given the EU courts renewed opportunity to define the landscape.

The types of agreement involving IPRs that are frequently caught by regulations include the granting of licences for use of one or more types of IPR, and collaborative R&D when two or more enterprises decide to work together in a technical field of mutual interest. The IPR aspects include use of existing rights, and ownership and use of rights that will be generated during the collaboration.

For these types of arrangements, the EU has prepared sets of regulations that set out what is and what is not permissible under the rules on competition. These are known as block exemptions. The easiest way to avoid a problem in the EU, and the UK, is to ensure that all clauses of an agreement fall within the exemptions according to the guidelines provided by the authorities.

Otherwise, an agreement must be assessed against the general principles of Article 81. In the past, it was possible to obtain approval from the EU for specific agreements, but this was removed in 2004 though it remains in the UK in a limited form. It is now necessary, as it is in the rest of the world, to self-assess agreements against the body of legal precedent, regulation and guidance: after half a century of nurturing by the EU, the rules should be clear to understand.

The authorities pursue a proactive role in investigating agreements, particularly in areas of the economy where competition seems to be lacking. They can also act if a third party, such as an aggrieved competitor, asks them to do so. If the authorities do become formally involved, they have wide-ranging powers to request information, take statements and conduct inspections that may include the use of unannounced 'dawn raids'. In all cases, it is advisable to cooperate fully, otherwise penalties may apply.

In enforcing the rules, the authorities have strong powers. The EU is allowed to impose fines of up to 10 per cent of the previous year's turnover of an enterprise. It also has broad powers to make orders that request the termination of agreements, disposal of assets or performance of specific activities. The UK authorities have similar abilities.

8.3 Agreements between undertakings

8.3.1 Background

Agreements between undertakings are largely addressed by Article 81 of the EU Treaty (see Figure 8.1) through EU Regulation 1/2003, and by Chapter I of the UK Competition Act 1998. If an agreement has anti-competitive effects according to the criteria of Article 81(1), and is not exempted because of any benefits outweighing those effects as outlined in Article 81(3), then it is considered void according to Article 81(2). Void agreements are legally unenforceable. The undertakings involved in such agreements may also be penalised or fined.

Agreements are more likely to infringe the rules on competition if they are between enterprises at the same levels of production, such as between manufacturers of similar products. These enterprises are considered natural competitors, and horizontal agreements between them are more likely to be restrictive of competition. Vertical agreements, such as one between a manufacturer and a distributor, are typically less restrictive of competition.

For example, it is recognised that appointing an exclusive distributor in a country is a very effective way of introducing a new high-technology product, and of overcoming differences of language and culture. It is also permissible for small or medium size enterprises in different countries to agree that each will make part of a product range and distribute the whole range in their respective home markets – these beneficial economies of scale are regarded as ultimately bringing benefits to the consumer.

8.3.2 Conditions for an agreement to be anti-competitive

To be deemed anti-competitive, an agreement must meet the four conditions of Article 81(1), as reflected in Chapter I of the UK Competition Act:

1. It is an agreement between undertakings, including decisions by associations of undertakings (e.g. trade unions, cooperatives).
2. It has an effect on trade, within or between Member States.

Treaty Establishing the European Union
Part Three - Community policies
Title VI - Common rules on competition, taxation and approximation of laws
Chapter 1 - Rules on competition
Section 1 - Rules applying to undertakings
Article 81

1. The following shall be prohibited as incompatible with the common market: all agreements between undertakings, decisions by associations of undertakings and concerted practices which may affect trade between Member States and which have as their object or effect the prevention, restriction or distortion of competition within the common market, and in particular those which:
   a. directly or indirectly fix purchase or selling prices or any other trading conditions;
   b. limit or control production, markets, technical development, or investment;
   c. share markets or sources of supply;
   d. apply dissimilar conditions to equivalent transactions with other trading parties, thereby placing them at a competitive disadvantage;
   e. make the conclusion of contracts subject to acceptance by the other parties of supplementary obligations which, by their nature or according to commercial usage, have no connection with the subject of such contracts.
2. Any agreements or decisions prohibited pursuant to this Article shall be automatically void.
3. The provisions of paragraph 1 may, however, be declared inapplicable in the case of:
   - any agreement or category of agreements between undertakings;
   - any decision or category of decisions by associations of undertakings;
   - any concerted practice or category of concerted practices,

   which contributes to improving the production or distribution of goods or to promoting technical or economic progress, while allowing consumers a fair share of the resulting benefit, and which does not:
   a. impose on the undertakings concerned restrictions which are not indispensable to the attainment of these objectives;
   b. afford such undertakings the possibility of eliminating competition in respect of a substantial part of the products in question.

Figure 8.1 EU Treaty Article 81 addressing anti-competitive agreements between undertakings

3. It has an effect on competition, such as prevention, restriction or distortion.
4. It has an appreciable effect, rather than being inconsequential.

Each condition will now be looked at in detail.

8.3.2.1 An agreement between undertakings

The term 'undertaking' is interpreted broadly. It includes enterprises, public organisations, partnerships, charities, trade unions, sole traders, inventors, consultants and designers. A parent in control of its subsidiary is regarded as a single undertaking. In a case involving Decca Navigator Systems, the subsidiary company,

Racal-Decca, along with the whole of the parent group, Racal, was considered a single 'undertaking'.

The 'agreement' need not be formal or even written; oral agreements are included, as are 'concerted practices' which cover any type of direct or indirect cooperation by two or more separate undertakings. When all the ten companies that among them make 80 per cent of all dyestuffs in the EU raised their prices by the same amount, at the same time, in three different years, this was a concerted practice. The fact that there was no evidence of collusion was irrelevant for the finding that Article 81 was infringed.

The 'decisions by associations of undertakings' include any decisions, regulations or recommendations by trade associations, standards councils and other representative bodies. SCK, an association of crane-hire firms in the Netherlands, operated a certification system and certain prohibitions on hiring that closed access to firms from other Member States. This was found to infringe Article 81 as it considerably disturbed competition for crane-hire in the single market. SCK was fined €300 000.

8.3.2.2 An effect on trade

There must be an actual or foreseeable, whether direct or indirect, effect on trade among Member States for the EU rules on competition to apply.

Grundig, the German manufacturer of dictating machines, appointed the Consten company as its exclusive dealer in France, and allowed it to be the owner in France of Grundig's trade mark Gint. Grundig prohibited its dealers in other countries from exporting to France. A German company bought Grundig machines in Germany and sold them in France. Consten sued for trade mark infringement. The decision was that the agreement between Grundig and Consten was anti-competitive and that the market should not be partitioned in this way.

Actions entirely within one Member State may also have an effect on interstate trade. A Dutch trade association of cement dealers that engaged in price-fixing was held to infringe Article 81(1) because the effect extended across the entire Member State and thus reinforced the compartmentalisation of markets on a national basis.

The UK rules on competition will apply when there is no appreciable effect on EU interstate trade. It is recognised that there will be borderline cases where it may be difficult to determine the applicable regime, so protocols have been put into place.

8.3.2.3 An effect on competition

The 'object or effect' of 'prevention, restriction or distortion of competition' must exist. This catches agreements having an anti-competitive effect, even if they do not intend to. Those with the 'object' are the most severely anti-competitive and attract the most punishment. These issues are determined by defining and analysing the relative market, which can be an onerous task.

The restrictive terms in a franchising agreement by Pronuptia, a retailer in bridalwear, were found to be acceptable as they protected the IPRs of the franchisor and the quality and integrity of the franchise system. This was beneficial to competition, not detrimental.

8.3.2.4 An appreciable effect

The effects must be appreciable in size according to the *de minimus* doctrine. The EU authorities have produced a notice indicating relevant market thresholds of 10 per cent for horizontal agreements, 15 per cent for vertical agreements and 10 per cent where it is unclear. The UK authorities apply a threshold of 25 per cent. Both authorities reduce the threshold to 5 per cent in cases where there are several companies involved in parallel networks of similar agreements.

A German producer of washing machines guaranteed territorial protection to exclusive distributors in Belgium and Luxembourg, which would usually infringe Article 81, but did not because it had a market share of less than 1 per cent.

The block exemptions often raise these thresholds in specific circumstances that are generally less anti-competitive.

8.3.3 Exemptions for beneficial agreements

If an agreement can be shown to have results that are beneficial to the economy, then the infringement of Article 81(1) may be exempted by Article 81(3) (see Figure 8.1). For example, a restrictive patent or know-how licence may be necessary to secure funding for commercialisation that will eventually result in technical or economic progress or improve the production or distribution of goods. The argument that the consumer will benefit if the agreement is implemented is a powerful one.

This was the case when GEC and Weir agreed in 1977 to pool their knowledge relating to the sodium circulators used to cool nuclear reactors. GEC was knowledgeable about nuclear reactors and Weir was knowledgeable about pumps and hydraulic systems. The agreement was found to be anti-competitive because it restricted competition between GEC and Weir and might affect trade between Member States in sodium circulators. This was particularly likely because there was only one customer in the UK – the UKAEA – so other common market companies would find it harder to sell in the UK. But the agreement promoted technical process and contributed to improved production and distribution of the goods. Each party became stronger and more knowledgeable in the others' technical expertise, and the consumer (interpreted broadly) benefited because liquid metal technology was improved.

The criteria for exemption are described in very broad terms in both Article 81(3) and the UK Competition Act. The authorities apply this breadth to specific agreements by using the practice of individual or block exemptions.

8.3.3.1 Individual exemptions

Before May 2004, the competition authorities could individually assess agreements in two ways. An undertaking could notify the authorities of an agreement to request negative clearance of Article 81(1), which confirmed that the agreement was not anti-competitive in nature. If the agreement fell within Article 81(1) then an undertaking could apply to the authorities for an individual exemption of Article 81(3) by justifying the beneficial effects.

This practice provided an undertaking with legal certainty, but was time consuming for the authorities, not scalable in light of EU enlargement and not present in other competition systems, such as the USA and Australia. With the practice removed, an agreement must now be self-assessed against the body of legal precedent, regulation and guidance.

8.3.3.2 Block exemptions

The criteria of Article 81 are described in very general terms, yet in reality many agreements are of similar and specific types. To address these common types of agreements, the authorities have developed a practice of producing regulations referred to as 'block exemptions'. The authorities also provide a considerable body of supporting, yet not legally binding, notices and material for further guidance in the application of the exemptions.

With the removal of individual assessment, block exemptions are now the primary means for an undertaking to design, adjust and self-assess an agreement against Article 81. Reflecting this change and its importance, the authorities have thoroughly reviewed, revised and modernised the practice of block exemptions in the last decade.

The block exemptions relevant to the IPR agreements will now be considered.

8.3.4 Exemptions for technology transfer agreements

8.3.4.1 Overview

The owner of a patent is not always able to exploit the invention themselves. The owner may be a small enterprise, or an enterprise that is not active in all possible fields of application, or may not even be a manufacturer at all. In these and other cases, the owner may want to grant a licence to a third party. The principles are considered in Chapter 9.

When granting such licences, the most satisfactory arrangement for both parties is if the licence is exclusive, that is, the patent owner agrees neither to use the invention nor to license it to any other parties. The licensee then has a market free of competitors, and the owner can negotiate higher royalty payments in return. The licensee will also have an incentive to invest in setting-up manufacturing and distribution systems.

An exclusive licence by definition excludes competition. The EU decided in the early 1970s that an exclusive licence covering a whole country was an infringement of Article 81(1), but also realised that innovation must be encouraged. There may have been considerable investment in the R&D that led to the invention, and there may be considerable value in introducing the technology into other parts of the common market.

The numerous issues of Article 81 relevant to typical technology transfer have been clarified by the EU in Regulation 772/2004 and its supporting guidelines. The regulation offers exemptions for agreements involving patents, designs, know-how and software copyright, but not trade marks or other IPRs, however in practice, many of the same principles are followed.

8.3.4.2 Conditions

The exemption applies where the combined relevant product or technology market share of the undertakings does not exceed 20 per cent. The threshold increases to 30 per cent if they are not competitors in the market. The exemption is lost if the market shares, subsequently, increase beyond these thresholds.

The agreement must be between two undertakings, and for the purpose of manufacturing or providing goods or services incorporating the licensed technology. The licensed technology must be the main object of the agreement, not just an ancillary part of it, and may be any combination of patents, know-how or software copyright. If the IPR is ancillary, then the R&D or specialisation block exemptions may be appropriate (see Sections 8.3.5.3 and 8.3.6.2).

'Patents' are defined to include similar forms of protection, such as topologies of semiconductor products and designs, including member state and Community Design rights.

'Know-how' means 'a package of non-patented practical information, resulting from experience and testing, which is secret, substantial and identified', which are often trade secrets necessary to work the patent:

1. '*Secret*': this does not mean totally unknown except to the licensor, but the information should not be generally known or easily accessible. It is the sort of information that benefits licensees by giving them a head start over competitors.
2. '*Substantial*': this requires that the information is necessary for the production of the contract goods.
3. '*Identified*': this means that the information should be capable of being described in some way so that it can be verified. It should be recorded, so if the information can only be transferred by demonstration (when the word 'show-how' is used), then the nature of the demonstration, its time and date, and the personnel involved should be recorded.

The current regulation contains a 'black list' and a 'grey list' of terms that are not acceptable. If any 'black list' terms are in the agreement, then the entire agreement is void, but 'grey list' terms only void themselves. The previous regulation contained a 'white list' of terms that are acceptable, but these are now part of the non-legally binding guidelines.

The exemption applies only for as long as the IPR has not expired or been declared invalid, or in the case of know-how remains secret (other than as a result of the licensee).

The guidelines contain a number of examples showing how the regulation applies to realistic commercial licensing scenarios.

8.3.4.3 Allowances

The guidelines to the regulation suggest that the following terms generally do not inhibit competition, and thus are allowable in licensing agreements.

1. *Confidentiality*: the recipient of the licensed technology can be obliged to keep it secret, even after the licence agreement expires.

2. *No subcontract*: the licensee can be prohibited from sub-licensing.
3. *Use after termination*: the licensee can be prohibited from using the licensed technology after the agreement terminates, provided the technology, whether patent, know-how or otherwise, remains valid.
4. *Infringements*: the licensee can be obliged to assist the licensor in enforcing the licensed IPR, which may include informing about misuse, or participating in litigation.
5. *Minimum royalty*: minimum royalty payments or minimum production requirements are acceptable.
6. *Marking*: it is acceptable to insist that the licensed technology is marked with the licensor's mark or name.
7. *Continuing royalties*: to help recover reasonable costs, it is allowable to extend royalty obligations beyond the validity of the licensed IPR, even if know-how has become public knowledge or patents have expired.
8. *Market allocation*: the use of the licensed technology can generally be limited to a particular field of application or product market, although some restrictions will apply.
9. *Quality*: it is acceptable to include requirements, such as tie-in obligations (which require that the licensee obtain other products and services from the licensor) to ensure that quality standards are met.

8.3.4.4 Restrictions

The following terms are considered to inhibit competition, and the exemption will not apply to licensing agreements that contain them directly or indirectly.

1. *Price restrictions*: no party to the licence can be restricted in its ability to determine prices for selling products to third parties.
2. *Output limitations*: there can be no limits on outputs or sales, other than those imposed upon the licensee in a non-reciprocal agreement.
3. *Market allocation*: allocation of markets or customers can only include:
   (a) *field of use*: requirements for the exploitation of the licensed technology to one or more product markets or technical fields;
   (b) *territorial*: the allocation of active sales (where the licensee actively advertises and solicits business) or passive sales (where the licensee merely responds to unsolicited orders) between parties, if the agreement is non-reciprocal;
   (c) *captive use*: requirements for the licensee to manufacture or provide the products only for its own use, including spare parts, or, in a non-reciprocal agreement, as a second source for a customer.
4. *No competition*: there can be no restrictions on the licensee's use of its own existing technology, or its ability to undertake any other R&D, unless the latter is necessary to prevent disclosure of licensed know-how to third parties.

In addition, if the parties are not competitors, then the following also apply:

5. *Price restrictions*: it is allowable to impose maximum or recommended sale price, but it is not a fixed or minimum price.
6. *Market allocation*: allocation of markets or customers can also include:

   (a) *territorial*: restrictions on passive sales into a territory reserved to the licensor or another licensee (only for a minimum of 2 years);
   (b) *wholesalers*: restrictions on sales to end-users;
   (c) *distributors*: restrictions on sales to unauthorised distributors;
   (d) *retailers*: restrictions on operating from unauthorised places.

If any of the following terms are present in a licensing agreement, then they do not prevent the agreement from being exempted, but the terms will be unenforceable:

7. *Grantback*: obligations on a licensee to exclusively license or assign any improvements or new applications of the licensed technology.
8. *No challenge*: an obligation not to challenge the validity of the licensor's intellectual property, including the secrecy of the know-how, although provision can be made for termination if such a challenge is made.

8.3.5 Exemptions for R&D agreements

8.3.5.1 Overview

The cost of R&D can be a heavy burden on an enterprise and in some areas of technology can be almost prohibitive. An effective commercial solution is for two or more enterprises with a common interest to work jointly, combining their resources, such as finance, assets or personnel. Often a formal joint venture (JV) is set-up, but not always.

Working together in this way can be anti-competitive. The parties are unlikely to compete with the JV, so if the JV parent enterprises are actual or potential competitors, then competition has been restricted by the JV agreement and Article 81(1) probably applies.

However, without working jointly, the R&D may not be done at all and consumers will never experience any potential benefits. Similarly, if the joint working is permissible in theory, but the parties are severely restricted as to how they can exploit the results, the work may not be carried out in the first place.

A typical situation regarding JVs occurred in the mid-1980s in the field of chemical engineering. BP owned rights in a catalyst used in the production of ammonia; Kellogg was a manufacturer of chemical engineering plant. The two enterprises agreed to the joint development of the chemical engineering plant to run the process using the BP catalyst. This is a classic case of an agreement that restricts competition and therefore infringes Article 81(1). However, the agreement was exempted under Article 81(3) because neither firm could have undertaken the work alone, and the restrictions in the agreement were regarded as reasonable.

Assistance is provided by EU Regulation 2659/2000, which offers a block exemption for qualifying joint R&D agreements. It lists what are known as white clauses, which are acceptable conditions, and black clauses: if an agreement contains a black clause it is automatically invalid and unenforceable. The regulations also set out some clauses which are frequently found in JV agreements which would not infringe the competition rules, but which are included for clarity. In this way, the regulation is used as guidance when constructing such agreements.

Further help is found in the EU notice that provides guidelines on horizontal cooperation agreements.

8.3.5.2 Circumstances

The form of the agreement to carry out joint R&D is irrelevant. The parties to it can set-up a separate JV company, in which each hold shares, or each party can allocate resources of money or personnel, either for R&D or for production. The parties can jointly place subcontracted work on a third party. Even collaboration in transferring know-how or in licensing IPRs can be regarded as a JV. The parties may agree that one enterprise does the R&D, and the other sells the resulting product. The format does not matter if the result is joint development: it is the effect of the competition rules that must be considered.

A JV agreement is considered broadly in the context of the entire EU. This occurred in connection with four JVs in the field of optical fibres in the 1980s. Corning Glass had patents on optical fibres and set-up a JV in each of the four countries with a local cable manufacturer. Since Corning was not a competitor in the cable market, and each cable manufacture was not a competitor in the optical fibre market, each JV itself was not anti-competitive. However, the market for cables was oligopolistic (i.e. there was a small number of suppliers, none dominant but all large) and the network formed by the four agreements was held to infringe Article 81(1). The agreements were exempted by reducing the agreed exclusive rights and removing the restriction on competition among the JVs.

In general, an agreement to cooperate merely on R&D at any stage up to industrial application will not be anti-competitive, unless the two parties are extremely large and dominant companies.

The risk of being found anti-competitive is greater in agreements that relate to exploiting the results of the cooperation. The parties will each wish to have a fair share of the results as far as exploitation rights are concerned, otherwise the JV would not be viable. The best way to achieve this without being anti-competitive is to use only the permissible, white clauses.

'Research and development' is defined broadly. It can cover both products and processes and includes any procedure by which knowledge is acquired, whether this is theoretical analysis or experimental study. The definition extends to experimental production, manufacture of prototypes, technical testing of the outcome and even establishing facilities for supply of the product or service. Applying for patents and other types of legal protection for the results of the R&D is also included. The line is drawn when a product is made or a process is used commercially, or know-how for such manufacture or use is communicated, or IPRs are assigned or licensed. This counts as exploitation, not R&D.

8.3.5.3 Conditions

For the exemption to apply to an R&D agreement, the regulation requires that a number of conditions be met:

1. *Defined scope*: there must be a programme for the joint R&D that defines the objectives and field of work.
2. *Pooled results*: all parties to the JV must have access to the results and be free to independently exploit the results using any necessary pre-existing know-how. However, suppliers of commercial R&D services, such as research institutes, may restrict their use of results to further research purposes only and non-competing parties may agree to limit exploitation to particular technical fields.
3. *Joint exploitation*: any joint exploitation must relate to results that are protected by IPRs, such as patent rights, or as secret know-how. The results must be decisive for the manufacture of the contract product or the application of the contract processes. It may be difficult to determine this before obtaining the results.
4. *Unrestricted supply*: if the agreement does not involve joint distribution, then any manufacturing parties must fulfil orders for supplies from all parties.

The duration of the exemption is limited according to how dominant the parties are in the market for the products resulting from the agreement. This ensures that a successful activity does not itself result in anti-competitive dominance.

If the parties to the agreement are not competitors, then the exemption applies for the length of the agreement. If the results are jointly exploited, then it continues for a further 7 years after the resulting products are first put onto the market, and will continue to apply while the combined relevant market share of the parties does not exceed 25 per cent. Having exceeded 25 per cent, then the exemption only applies for another 1 or 2 years. For competing parties, the exemption applies only while the 25 per cent threshold is not exceeded.

The experimental use of a product, even by a customer, does not count as 'putting products on the market': exploitation is regarded as beginning at a later stage.

8.3.5.4 Restrictions

The following terms are considered to inhibit competition, and the regulation will not exempt R&D agreements that contain them directly or indirectly.

1. *No competition*: restriction of independent R&D is not allowed in an unconnected field during the agreement, or in either a connected or unconnected field after completion of the agreement.
2. *No challenge*: it is not permissible to prohibit challenging the validity of any IPR relevant to the agreement, including those resulting from the agreement, however it is allowable to provide for termination upon such a challenge.

3. *Output or sales restrictions*: these cannot be present.
4. *Pricing restrictions*: the parties cannot fix the prices to be charged to third parties for the resulting products.
5. *Customer restrictions*: there can be no limits to the customers that can be served after 7 years from the time the resulting products are put onto the market.
6. *Passive sales prohibitions*: are not allowed in territories even if they are reserved for other parties.
7. *Active sales prohibitions*: are not allowed in territories reserved to other parties after 7 years from the time the resulting products are first put onto the market.
8. *Licensing restrictions*: the parties must not be prohibited from granting licences to third parties to use the results of the R&D programme if exploitation by one party is not provided for or does not take place.
9. *Refusal to supply*: it is not allowed to restrict meeting the demand of users or resellers who would market the resulting products in other Member States.

However, these prohibitions do not apply to setting production targets during joint exploitation, or setting sales targets and fixing prices for immediate customers during joint distribution.

8.3.6 Exemptions for specialisation agreements

8.3.6.1 Overview

Enterprises sometimes agree to specialise in producing goods, and IPRs often form part of these agreements. The usual case is where a manufacturer decides to outsource production, or where two manufacturers agree to a joint production.

A more complex case is where two manufacturers obtain and rebadge products from each other to supplement gaps in their product portfolios so as to present a complete offering to their customers. This strategy has been used in the automobile industry.

The manufacturers might be competitors in the same sector generally, but not in exactly the same products. There is considerable scope for anti-competitive behaviour, but at the same time, the consumer may benefit from the lower costs of outsourcing, or the single source of an entire product range.

To specify the acceptable arrangements, EU Regulation 2658/2000 offers a block exemption for certain types of specialisation agreements, and further help is found in the EU notice that provides guidelines on horizontal cooperation agreements. This exemption only applies where IPRs are not the primary object of the agreement, otherwise the technology transfer block exemption in EU Regulation 772/2004 may apply.

8.3.6.2 Conditions

The exemption applies only if the combined relevant market share of the parties does not exceed 20 per cent, and should it increase beyond 20 per cent, then the exemption will only apply for another 1 or 2 years.

Three types of agreements are exempted:

1. *Unilateral specialisation agreement*: where one party stops producing certain products and agrees to purchase them from a competitor who agrees to produce and supply them.
2. *Reciprocal specialisation agreements*: where two or more parties reciprocally stop producing certain but different products and agree to purchase them from the other parties who agree to supply them.
3. *Joint production agreements*: where two or more parties agree to jointly manufacture certain products.

The parties are only allowed to engage in exclusive purchase or supply arrangements with each other. Also, if they do not sell the products independently as part of a joint production agreement, then they can arrange for joint distribution through a non-competing third party.

The parties are neither allowed to fix prices when selling to third parties, nor allowed to limit outputs or sales, or allocate markets or customers. They can agree upon capacities and volumes of production or exchange between themselves, or set sales targets and fix prices for immediate customers of a joint distributor.

8.3.7 Exemptions for vertical supply and distribution agreements

8.3.7.1 Overview

Enterprises frequently engage in agreements with each other at different levels of the production or distribution chain, such as for the supply of raw materials, or the distribution of finished products. It is common to find IPRs playing a supporting role in these agreements, such as a patented production process or a trade mark used as part of advertising material.

There is less scope for anti-competitive behaviour in vertical agreements than with horizontal agreements because the parties are usually not natural competitors as they are at different levels in the economy. However, there are still opportunities for restrictive practices in the terms of the agreement.

EU Regulation 2790/1999 provides a block exemption for vertical agreements concerning the supply and distribution of goods or services. It only applies to IPRs in as far as they are not the primary object of the agreement, but are directly related to fulfilling the primary objective of the agreement. Otherwise, the technology transfer block exemption in EU Regulation 772/2004 may apply.

Further help is found in the EU notice that provides guidelines on vertical restraints.

8.3.7.2 Conditions

The exemption applies to agreements between undertakings that operate at different levels of the production or distribution chain, at least for the purposes of the agreement, and the agreement must relate to the sale or purchase of goods or services. Many vertical agreements do not require exemption in the first place, as they simply do not meet the criteria of Article 81(1).

The parties involved in the agreement must not have more than a 30 per cent share of their relevant market, as either suppliers or buyers. If the exemption does apply, but their market share increases beyond 30 per cent, then the exemption will only apply for another 1 to 2 years.

If the parties are competitors, then the exemption only applies if the agreement is non-reciprocal. If they do not compete directly in the manufacture of the goods or in the provision of the services, the buyer must also not have a turnover greater than €100 million.

8.3.7.3 Restrictions

The exemption will not apply to entire agreements that contain any of the following restraints, either directly or indirectly.

1. *Price restrictions*: although the supplier can impose a maximum or recommended sale price, it cannot in any way fix or impose a minimum sale price.
2. *Territorial restrictions*: passive sales in response to unsolicited orders can never be restricted, although active sales to a specific territory can, but only if that territory has been allocated to the supplier or another buyer.
3. *Selective distribution*: the distributors in a selective distribution system cannot be restricted in who they may sell to (including unauthorised distributors), however, they can be required to sell from only one given location, but they must be free to trade the contract goods with other distributors, not just the supplier.
4. *Replacement parts*: the supplier cannot be restricted in sales of spare parts to end-users, independent repairers or parties other than the distributors.

If the following restrictions are present in the agreement, the exemption will not apply to them, but may apply to other parts of the agreement.

5. *No competition*: restrictions on the buyer to purchase 80 per cent or more of their total requirements cannot exceed 5 years, unless the buyer operates from premises owned by the supplier.
6. *Post-term no-competition*: upon termination of the agreement, the non-competition obligation is allowed only if is necessary to protect know-how, and is limited to the premises from which the goods were sold, and to a maximum of 1 year.
7. *Selective distribution*: the supplier cannot prevent dealers from selling competing goods.

8.3.8 Exemptions for subcontracting agreements

An EU notice states that subcontracting agreements do not fall foul of Article 81 and are to be distinguished from technology transfer agreements as covered by EU Regulation 772/2004.

Where the contractor provides equipment or technology as part of the agreement, it can require that any use, results or disclosure of such materials be only for the purpose of the agreement. It is made clear that this applies to all forms of IPRs, whether by themselves as patents, design rights, copyrights, confidential information, trade secrets, or fixed in works, such as studies, plans, documents, dies, patterns and tools.

However, these restrictions cannot apply to any general knowledge that the subcontractor has or could otherwise obtain, nor to any secret information that otherwise becomes public knowledge. If this were not prohibited, a contractor could unfairly stifle the independent activities of the subcontractor.

The subcontractor can be required to provide the contractor with non-exclusive use of any improvements or inventions that are devised.

8.4 Dominance of undertakings

8.4.1 Background

The dominance of undertakings is largely addressed by Article 82 (see Figure 8.2) of the EU Treaty, implemented through EU Regulation 139/2004, along with Chapter II of the UK Competition Act 1998, the UK Enterprise Act 2003 and remnants of the UK Fair Trading Act 1973. EU and UK legislation and practice are now harmonised.

The existence of a dominant position itself is not prohibited, merely its abuse: particularly those hardcore abuses listed in Article 82.

Treaty Establishing the European Union
Part Three - Community policies
Title VI - Common rules on competition, taxation and approximation of laws
Chapter 1 - Rules on competition
Section 1 - Rules applying to undertakings

Article 82

Any abuse by one or more undertakings of a dominant position within the common market or in a substantial part of it shall be prohibited as incompatible with the common market in so far as it may affect trade between Member States.

Such abuse may, in particular, consist in:

a) directly or indirectly imposing unfair purchase or selling prices or other unfair trading conditions;

b) limiting production, markets or technical development to the prejudice of consumers;

c) applying dissimilar conditions to equivalent transactions with other trading parties, thereby placing them at a competitive disadvantage;

d) making the conclusion of contracts subject to acceptance by the other parties of supplementary obligations which, by their nature or according to commercial usage, have no connection with the subject of such contracts.

Figure 8.2 EU Treaty Article 82 addressing abuse of a dominant position by undertakings

By definition, dominant undertakings lack effective competition. Without this competition, it is easier for those undertakings to charge unnecessarily high prices or to engage in other abusive practices. For these reasons, such undertakings require special attention to address both intentional and unintentional effects of their dominance.

A 'dominant position' is determined in the context of the relevant product and geographic market which may not be easy to define. In a case involving Continental Can, it was argued that the relevant market was light cans for preserving fish but the final decision was that the market was tin cans for any type of food.

In the context of the relevant market, dominance is established by numerous criteria, including market share, barriers to entry, vertical integration, distribution arrangements, conduct, performance and ownership of key resources – including IPRs. The authorities have stated that dominance cannot be ruled out even if the market share is as low as 20 per cent if other factors are present. Usually, a market share of at least 40–45 per cent is required.

The authorities have produced a number of notices to clarify how the relevant market is defined and how concepts, such as undertakings, dominance and turnover, are construed. These are vital in clarifying the rules, as there has been much confusion and concern, as evidenced by numerous cases.

It is important to understand that dominance does not just apply to large enterprises. It can apply to a small business, if that small business has a grip on a niche market, perhaps by owning key technologies, a unique presence on the Internet, or a successful product or service. It is possible for purchasers to have dominance, just as much as suppliers: supermarkets are an example.

The EU courts have made it clear that Article 82 cannot be applied for ownership of IPRs *per se*, but can be applied to their abusive use. For example, a patent provides monopoly rights by its nature, and a holder may legitimately refuse to license.

The rules do not prevent the exercise of other legitimate rights, irrespective of motivation. In a case in the UK courts involving electrostatic powder coatings, it was held not to be an abuse of a dominant position to intend to drive a competitor out of business by bringing legal proceedings, as all companies have the right to go to court.

For the EU rules on competition to apply, the dominance must affect a substantial part of the EU market. Otherwise, only the UK rules on competition can apply. There is increasingly little difference between the regimes, but there can be confusion about which is applicable in borderline cases. The authorities have produced protocol and guidance to help.

8.4.2 Application

If an undertaking owns an essential facility, traditionally a seaport or an airport, for example, then refusal to permit access on reasonable terms may be abusive. This principle has been applied to IPRs. In what seems to have been an otherwise exceptional case, the EU courts found that refusal of three television companies in the UK and Ireland to provide reasonable licences for copyright in their TV listings for a TV guide was not justifiable. It was found that the TV guide was beneficial for the consumer, and licences were ordered.

Restricting the availability of replacement or consumable parts of a product, including the ability of competitors to produce compatible parts, can be abusive. Third parties claimed that it was an abuse of a dominant position for Volvo not to grant licences for the production of spare parts. The EU courts did not answer but clearly stated that Volvo might be guilty if it had refused to supply spare parts to independent repairers, charged unfair high prices or refused to produce spare parts for models still in circulation.

Hilti faced a similar problem in relation to replacement nail cartridges for its nail guns. It infringed Article 82 by demanding excessive royalties and refusing to supply cartridges to competing producers of nails, among other discount and tie-in abuses designed to close-off competition. Hilti argued that some of these restrictions were necessary for quality control reasons, but the EU courts did not fully agree.

In these cases, compulsory licences were ordered or the undertakings were required to stop their restrictive practices. Another approach is to request that products are disclosed or unbundled (i.e. where a product is split and sold in separate parts). The EU directed the unbundling of the telecommunications local loop as a way to improve competition in the sector. More recently, the authorities ordered that Microsoft unbundle the media player component from its Windows operating system and release details of its software interfaces to aid competitors in providing interoperable products.

Buying an enterprise that has been granted an exclusive licence, so as to strengthen the purchaser's market position, can also be a problem. For example, Tetrapak, a very large supplier of machinery and cartons for packaging milk, bought Liquipak, the exclusive licensee of a patented system for sterilising milk cartons. This was found to be contrary to Article 82 as Tetrapak already had a substantial technological lead over its competitors, and the acquisition would have closed what is otherwise a large, valuable and beneficial market.

It is clear that dominant companies have a special duty to avoid restrictive practices. This involves an understanding of the relevant laws, markets and technologies.

8.5 Free movement of goods and the exhaustion of rights

It is a fundamental principle of the EU, as specified in Articles 28–30 of the EU Treaty (see Figure 8.3), that goods are to have free movement between Member States. There are often conflicts between this principle and the existence of current or prior national laws and systems, especially as they relate to IPRs. Article 30 does permit a limited exception to this principle in accounting for the inherent nature of IPRs, such as with compulsory licensing which is required in some Member States.

A community-wide exhaustion of rights doctrine has been developed for IPRs by the EU courts. This allows an EU-based IPR owner to benefit from protection once only, so that in the sale of goods protected by patent or design rights, the owner cannot use those rights to stop any further movement of the goods once they have been put onto the market in the community. The purchaser is then able freely, and commercially,

Treaty Establishing the European Union
Part Three - Community policies
Title I - Free movement of goods
Chapter 2 - Prohibition of quantitative restrictions between Member States
Section 1 - Rules applying to undertakings

Article 28

Quantitative restrictions on imports and all measures having equivalent effect shall be prohibited between Member States.

Article 29

Quantitative restrictions on exports, and all measures having equivalent effect, shall be prohibited between Member States.

Article 30

The provisions of Articles 28 and 29 shall not preclude prohibitions or restrictions on imports, exports or goods in transit justified on grounds of public morality, public policy or public security; the protection of health and life of humans, animals or plants; the protection of national treasures possessing artistic, historic or archaeological value; or the protection of industrial and commercial property. Such prohibitions or restrictions shall not, however, constitute a means of arbitrary discrimination or a disguised restriction on trade between Member States.

Figure 8.3 EU Treaty Articles 28–30 addressing restrictions on free movement of goods by undertakings

to sell the goods in another Member State, perhaps where the original prices are higher. The principle also applies where an IPR owner has permitted manufacture by a third party under a licence.

In an early case, Deutsche Grammophon was not able to rely upon German copyright law and object to the importation of records into Germany that were originally sold by its subsidiary in France. This was later qualified to distinguish between the sale of goods and a specific performance of a service, in a situation where the EU courts did not allow Cine Vog's films, originally transmitted in France, to be retransmitted by another party into Belgium from Germany.

The EU courts have developed a form of international exhaustion of rights by giving IPR owners the ability to prevent reimportation of goods put onto the market outside of the EU, thus reinforcing the notion of a 'Fortress Europe'. In the Silhouette case, an Austrian trade mark owner was able to prevent the reimportation of spectacle frames that it had originally sold in Bulgaria. Recently, the online CD music retailer CD-Wow reached an out-of-court agreement with the British Phonographic Institute to stop importing unauthorised CDs from outside the EU. This rule has been controversial, especially in the eyes of consumers.

To prevent disguised restrictions on trade, the EU Treaty prohibits any 'measures having equivalent effect'. The application of this has frequently occurred when goods have been branded or packaged differently across Member States. The EU courts have allowed parallel importers to apply local trade marks to identify the goods, and to even relabel or repackage the goods. In one case, Centrafarm was allowed to purchase drugs with the 'Serenid' trade mark in the UK, and resell them under the local 'Seresta' trade mark in Holland despite an objection by the owner of the marks, American Home Products. The EU courts saw through this artificial market partitioning.

The EU courts have established that some restrictions on free movement are allowed when necessary for the specific subject matter of IPRs, or where consent was not involved. Pharamon was allowed to prevent the import of its patented products into a Member State as the products had been put onto the market under a compulsory patent licence in another Member State. It was irrelevant that Pharamon had accepted royalties as compensation for the licence.

In other countries, there are similar principles for IPRs. The USA refers to exhaustion of rights as the 'first sale doctrine', and although it applies in a more complex manner, there is no international exhaustion either. In Japan and Australia, the approach is the opposite: there is international exhaustion, meaning that IPR owners cannot use their rights to prevent reimportation of goods put on the market outside of these countries.

8.6 Summary

Intellectual property rights strike a balance by protecting investment without hindering the operation and progress of society. However, in their exercise, there is the scope for abusive and restrictive practices that can distort the competition that is necessary for a healthy and progressive economy. This has long been understood.

The development of the EU brought a focus to the principle of free movement of goods and resulted in a concerted effort to develop and enforce a set of rules on competition against any restrictions on trade between Member States. The result is a progressive and active network of competition practice across the community that has application to all forms of economic behaviour.

There is a comprehensive body of material for the rules on competition as they relate to IPRs, either published by the authorities or developed through the courts. It is essential to consider this material in all activities that involve IPRs, especially licensing agreements and collaborative arrangements with other companies. Failure to do so can be costly.

Chapter 9
Licensing and litigation

An intellectual property (IP) licence is effectively a promise not to sue for infringement of an intellectual property right (IPR). If litigation occurs, it is frequently settled by an alleged or proven infringer taking a licence. Payments of damages in litigation are often set on the basis of a notional royalty. It is therefore appropriate to consider the two topics in a single chapter.

9.1 Licensing IPRs

9.1.1 Why license?

While the most direct way of recovering the cost of developing an engineering technology is to make a product and sell it, with the IPRs keeping competitors out of the market, this is not the only way.

Sometimes an idea is so fundamental that one company cannot itself exploit it to the full. This was the position when the transistor was invented. Western Electric granted licences under the basic transistor patents to a large number of companies worldwide, who all paid a small royalty (1 or 2 per cent) for the right to use the invention.

In that case, the patent owner clearly had the industrial muscle to litigate if the patent was infringed, so taking a licence was the safest course. Some licences are granted after the early stages of litigation and this is often the best solution to the dispute, and the most profitable in the long term for both parties. It is reported that in the year 2000, IBM made US$1.5 billion from granting licences. Both large and small companies can benefit from exploitation of IPRs in this way.

There are other good reasons for granting a licence. A small company may not have the capacity to meet the full demand for a product, whether in the UK market or overseas. The solution is to allow others to use the IPRs in return for payment.

A patent may have a broad scope, but the owner may be interested only in one part of the field. An invention relating to electric motors generally may be devised in a company which produces only low power motors – the invention can be exploited fully by licensing it to another company which makes high-power motors, and which has the appropriate sales network already in place.

A manufacturing company may take a business decision to move away from a particular area of engineering, but its engineers will retain their skill and expertise; these could be made available to others in a know-how licence lasting a short period while the expertise remains up to date.

An IPR may be owned not by a manufacturer but by a university or research association. The options for seeking a return on the R&D investment are to license or to assign. Assignment tends to be for a one-off payment and the value of IPRs is always extremely difficult to judge at the early stages of exploitation; a licensor has a greater chance of benefiting from a very successful exploitation by retaining a link and taking a percentage of the profit. On the other hand, the exploitation may be unsuccessful so taking the one-off payment may be advantageous.

Turning now to the advantages of taking a licence, a company can move into a new market quickly and save itself several years of development effort. The full details of an engineering process may be available, with IP protection to discourage others, so a company can take a licence and leapfrog its competitors, or enter an entirely new area of engineering.

The special arrangements for licensing use of computer programs are explained in Section 2.11.

9.1.2 What can be licensed?

Any IPR can be licensed individually, or a bundle of rights can be included; some licences involve all six rights giving very full technical and commercial protection. At an early stage, the relevant bundle of rights must be identified at the same time as the technology available for licensing is itself being defined.

If a product is already being manufactured, or a process used industrially, the technology is fully developed. In other cases, development may be much less advanced and it may be beneficial for the owner to spend a little time and money taking it further to give a more marketable package of legal rights.

If an IP licence is to be granted in another country, it can only include rights that exist in that country. Is there a patent there? If not, is it still early enough to file an application? Does copyright extend to manufacturing drawings in the country of interest? Is computer software specifically protected by copyright in that country, or is protection more speculative? Does the concept of confidentiality exist? Will unfair competition law give some protection?

Another important factor is whether the potential licensor envisages a long-term close relationship, beginning with initial transfer of know-how and continuing with exchange of information on improvements. Alternatively, is this to be a 'bare' patent licence with the owner simply promising not to sue in return for royalty payments?

Whatever the aim of the IP owner, these basic points should be considered as early as possible.

9.1.3 *Finding a licensee or licensor*

This section only applies to a licence granted by a willing licensor to a willing licensee on an arms-length basis, that is, the companies are not parent/subsidiary or connected by a joint venture (JV) agreement and there has been no threatened or actual litigation over the rights.

Multinational companies usually have little difficulty in identifying potential licensees from their existing engineering contacts worldwide. For smaller companies, finding a licensee may be more difficult. If a company has no information at all, it can advertise that the technology is available for licensing in various publications, such as trade journals. One sometimes sees advertisements for technology brokers who offer to bring potential licensees and licensors into contact. Most are reputable, a few are not, so care should be taken if this route is used.

When a potential licensee (or several) has been identified, the IP owner needs to consider if that company has the ability to manufacture the product, has the appropriate marketing and distribution capabilities and the expertise to provide the relevant services. Naturally, a good reputation is also important. Once a potential licensee has been identified, an approach can be made initially on a confidential basis. Setting up a good working relationship as soon as possible is very helpful; there will be hard bargaining in the future over details, but the ultimate aim should be that both parties feel they have benefited from the grant of a licence; a win–win situation.

In the reverse direction, larger companies often own patents which are not being actively exploited; the patent specifications are public literature and an approach to the patent department can lead to the grant of a licence to a small company which identifies an opportunity in this way. Patents which are held to keep a large competitor out of the market, 'defensive patents', may be available under licence to a small company with a niche market.

9.1.4 *Types of licence*

A licence can be exclusive, sole or non-exclusive.

If an IP owner grants an exclusive licence, the owner promises not to compete with the licensee and not to grant other licences in the same field. The licensee therefore has no competitors and must expect to pay for the privilege. The owner is also taking the risk that the licensee will not be successful, and it is usual to have performance clauses such as minimum royalty payments or a requirement to sell minimum quantities, with the licence being converted to non-exclusive if the minimum is not achieved.

A sole licence means that the owner is granting only one licence, but is itself going to compete in the same market. A non-exclusive licence means that other licences may well be granted and the owner may also be competing.

An IP licence can be granted orally but it is highly preferable if the agreement is in written form. This provides a good record of what was agreed initially; memories tend to fade with time and also tend towards a preferred position even if this was not the real position. As with all other legal agreements, a record is mainly important when things go wrong. This is the reason that legal advisors ask awkward questions when the parties are working together very happily in the initial stages.

9.1.5 Financial compensation

The type of payment and its level depend on what is being provided by the owner and what is being gained by the licensee. IP licences usually last for a considerable period and a good working relationship is best achieved if both sides are clear about the benefits and see the arrangements as fair in the long term.

If a licence is granted before a technology is fully developed, the potential value is difficult to judge. The IP owner may believe the technology will be extremely successful and profitable, the licensee may be less sure. The licensee may need to set up manufacturing facilities and a distribution network and it may be a substantial period, even years, before any benefit is detectable.

In general, payment is made in one of two ways or a combination of both; an initial lump sum and a continuing payment known as a royalty. Often the lump sum payment is made when confidential know-how is disclosed because the risk to the owner is high; information is largely irrecoverable. A royalty can be set as a percentage or as a fixed sum per item sold, the arrangements are almost infinitely variable.

Inevitably, taxes fall due on almost all payments. The type of tax depends on the type of IPR, the type of legal person receiving the payment and whether transfers to or from other countries are involved. Expert advice is essential to minimise the tax burden.

To judge the fairness of the payments, one needs to look at the whole package and the likely risks and benefits to both sides. Only then can the amount of money involved be fully evaluated. This follows the best negotiating practice; the two parties should discuss all aspects in detail first so that the package is fully defined and evaluated, and only then talk about money.

On this basis, several types of clause in a licence agreement are explained briefly below with the effect on the size of a payment mentioned where possible, and the subject of payment levels discussed afterwards.

9.1.6 Format of a licence agreement

9.1.6.1 The parties

The record of an IP licence agreement should begin by defining the parties. One needs to know if one is dealing with a subsidiary or its parent and the parties should both be legal entities, not a management group within a company. It is not essential to refer to the registered company; one merely needs to identify the party in full, by quoting the registered address or the trading address.

The relative sizes of the two parties will probably be clear. A large company always has an advantage in negotiation because it can bear the cost of litigation if agreement is not reached, and is therefore in a stronger negotiating position.

9.1.6.2 Recitals

These are the clauses beginning with 'whereas'; they set out the background to the licence and give a very brief summary of why the parties have reached agreement.

9.1.6.3 Definitions

A licence agreement is a legal document and the important terms should be identified and defined to minimise any arguments at a later date. A defined term is sometimes used in the body of the document with a capital first letter, but this is not always the case.

Patents, registered designs and registered trade marks may be defined by reference to a schedule at the end of the document, where the registration numbers, dates, titles etc. are given in full. Drawings and designs which are protected by copyright and design right can also be listed by an identifier.

9.1.6.4 Grant

This is usually a brief clause which sets out what the licensee can do, where it can be done, and for how long. The permission may extend to the full use of all IPRs or may be limited to a particular geographical or technical field. Suppose a patent relates to electric motors of any power; the licensee could be permitted to make electric motors of any power in any country in which a patent has been granted, or it could be limited to manufacturing in the UK for sale in the UK of electric motors up to a defined power.

This clause makes it clear whether the licence is exclusive, sole or non-exclusive; exclusive licences usually attract higher payments because there is no competition in the field. The grant can be for the life of the patent, or for a shorter period, renewable if the parties agree.

9.1.6.5 Know-how

A licence agreement will record whether there is to be transfer of know-how by provision of written records, by exchange of personnel or by demonstration. Often named employees will be involved and there may be a time limit, such as the named expert being available on signing of the agreement for a specified number of working days in the licensee's premises to give advice and assistance.

9.1.6.6 Improvements

There may be a clause under which the licensee is obliged to tell the licensor about improvements to the technology, and there may be a promise from the licensor to assist the licensee, which would justify a higher payment level.

9.1.6.7 Maintenance fees

Sometimes there is a clause setting out who is to pay renewal fees for patents etc.

9.1.6.8 Infringement

It is conventional to have a clause in which both parties agree to tell each other if they become aware of any infringements of the IPRs, and sometimes the licensee agrees to help with any litigation, frequently at the licensor's expense.

9.1.6.9 Use of trade marks

If the licence includes the right to use the licensor's trade marks, then provisions should be included which set out how the mark is to be used and in what format, such as size and colour, on the product, on packaging, manuals etc.

9.1.7 Royalty payments

Once both parties have identified what information is to be supplied, the scope of the permission to use the IPRs, and the future obligations of each side to the other, they can begin to discuss financial aspects. Preferably, the parties recognise the risks and benefits to both sides so that a win–win situation is reached. The licensor should recognise that a licensee may have a substantial start-up cost with investments in a new or modified production line and after-sales organisation. A promise of long-term assistance from an IP owner who will also bear the costs of patent prosecution and any litigation is obviously more valuable to a licensee than a simple permission to use a patent without supporting information.

Both sides will wish to make a profit, and it is often from expected profit that royalty rates are calculated. Very broadly, a reasonable royalty is often taken to be one-quarter or one-third of an anticipated gross profit, that is, before payment of tax or interest, or other royalty. Profit on a manufactured item is often about 20 per cent of the selling price so this would lead to a 5 per cent royalty, which is very often the agreed rate. There may also be an 'industry standard' royalty rate. However, for some products such as computer software, the production costs are minimal, that is, the cost of copying onto a disk, and most of the costs relate to the development; royalties of up to 50 per cent are not unknown in such circumstances. At the other end of the scale, products sold in high volume often carry much lower royalties, such as 1 or 2 per cent of the selling price of early transistors.

As an alternative to a percentage of selling price (or a notional selling price for transfer between group companies), a fixed sum per sale can be set. In addition, a royalty can decrease with time as sales volume increases, or can increase with time to take account of the start-up costs of the licensee and improve the chances of initial success by delaying substantial payments.

The arrangements are almost infinitely variable and rules of thumb can be ignored if the circumstances justify it and the parties agree.

Other factors supporting a high royalty rate are broad patent cover for the entire product; a patent which has been tested in court and held valid; the licensor's success in the home market and the associated reputation; or a high level of improvement in an engineered product or process. Royalties would be lower for a patent covering only a small part of the product, a small level of improvement or a licence with little in the way of support from the licensor.

9.1.8 Licensing professionals

Large companies with a substantial licensing activity often have specialist professionals with an engineering or scientific background in the appropriate technical field, and considerable negotiation skills. These skills are also offered by independent licensing consultants. Both types of professionals are often members of the Licensing Executives Society whose website is at www.les-europe.org or see the address at the end of the book. In companies, patent attorneys sometimes deal with licensing, and solicitors, whether in companies or in private practice, also negotiate and prepare licence agreements. Such professionals will also be familiar with the constraints of EU Competition Law (see Chapter 8) or equivalents such as anti-trust law in the USA or Japan.

9.1.9 Compulsory licences and licences of right

If a patent or a registered design is not being used in the UK, or not used as much as is feasible, an interested party can apply to the Comptroller of Patents for a compulsory licence and (if the patentee cannot provide a good defence) will be permitted to use the invention. A reasonable royalty rate is set. The patent must have been granted for longer than 3 years.

A patent owner can also arrange for a patent to be endorsed 'Licences for Right', which means that a licence will be granted automatically to any applicant. Patent renewal fees are halved if such an endorsement is made.

The licence of right provisions applying in the last 5 years of design right are explained in Section 3.2.8.1.

9.2 Litigation

9.2.1 Introduction to IPR litigation

Full litigation over infringement of IPRs is quite rare in the UK, but engineers often ask many questions about the process so an outline is given here.

In the UK, taking legal action is usually regarded as absolutely the last resort and all other routes are tried first to settle a disagreement and to maintain business relationships. This is not necessarily so in other countries, especially the USA, where bringing a legal action is regarded as a normal business option although an expensive one.

The costs are very high. Until recently it was an accepted saying that it was better not to sue for patent infringement because a company could not afford even to win the case, never mind lose it. In the last few years, there have been two changes which have reduced the costs. One is a change in procedure in the High Court which has resulted in trials becoming shorter, the other is the introduction of the Patent County Court. Both are described below.

Starting formal legal proceedings certainly concentrates the mind and shows that an IP owner is serious. The vast majority of IP cases are settled well before they come to court, as the parties both try to minimise legal costs. Negotiations to settle will often proceed at the same time as the formal case which can be stopped at any time, even at the door of the court, or part way through the trial.

In some cases, the legal position is clear from the beginning. If copyright material or a design has been copied, the copier was aware of doing so although proving this may be difficult for the owner. Often a trade mark has been deliberately selected to ride on the back of a more famous mark and the legal consequences are fairly obvious.

In other cases the outcome is not straightforward. Patent infringement involves construction of patent claims; is the defined invention novel and not obvious; is the claim infringed? Interpretation of the wording may lead to genuine differences of opinion, which may not be settled until the case proceeds for some time.

The tests for infringement of each type of IPR will be found in the chapter describing that right.

9.2.2 Benefits of winning a lawsuit

Intellectual property owners take infringers to court for two main reasons; to protect the market for their products by stopping competitors from selling, or to force another company to take a licence – this is often so when licence negotiations have broken down but a potential licensee clearly intends to use the IPR. Sometimes, there are several infringers and one is selected to make the point to the others. Whatever the reason, in successful litigation the expenditure of the owner on developing the innovation can sometimes be recouped, in the first case by higher direct sales, in the second by receipt of royalty payments. However, the most important factor is sending a message to others that the owner is willing to defend the IPR.

Either the owner of an IPR or the exclusive licensee of that right can begin proceedings for infringement; an exclusive licensee can suffer losses in exactly the same way as the owner, and the right to sue is automatic as far as patents, copyright and design rights are concerned (but not registered designs). In this section, 'owner' includes an exclusive licensee.

The remedies available in court allow the first aim of the owner to be achieved, that is, stopping a competitor from using the right, but do not directly cover the grant of a licence; it is up to the parties to reach an out of court settlement and agree terms, the court will not impose a licence.

The types of remedies which can be awarded are:

- an injunction against further infringement
- delivery-up of infringing goods
- damages or an account of profits
- a certificate of contested validity
- costs.

9.2.2.1 Injunction

A formal court order is made telling the infringer to stop the infringement. This remedy can be awarded after full trial, but in IP cases it is almost always ordered at an early stage when it is known as an interlocutory or interim injunction. This is one time when the court can act fast, for example, in a matter of days or even hours, provided the IP owner also acts fast. If the owner learns of infringement and goes to court immediately, within days or a few weeks at most, the court will act promptly. If the owner waits for many weeks or even months without good reason, the court will take the view that speed was not important to the owner and an interlocutory injunction will not be available; the owner must go to full trial.

To obtain an injunction, proving infringement is not in itself sufficient, other factors are also taken into account. A fairly obvious infringement with no obvious defence is the starting point, and any alleged infringement must be serious, not trivial. Another factor considered by the court is the 'balance of convenience'; if no court order is made, will the losses suffered by the IP owner be greater than those that would be suffered by the alleged infringer if an order is made and it later turns out that it should not have been made? Further, an interlocutory injunction will usually not be granted if the payment of damages in the long term will be sufficient to compensate the IP owner. This would be the case if the owner is not a manufacturer and can only gain benefit by granting licences; a payment is then sufficient in the circumstances to recompense the owner.

The court often insists on a 'cross undertaking in damages'. This is a promise by the IP owner to compensate the alleged infringer for its losses if it turns out that the injunction was not justified. A large company will usually be regarded as sufficiently stable to be able to make the promised payment at a later date, a smaller company may have to provide some form of security.

The court's decision relating to an interlocutory injunction, whether it is granted or not, is often the end of the matter. If the losing party decides not to take the argument further, there may never be a full trial.

9.2.2.2 Delivery-up or destruction

If an infringer has a stock of infringing goods, it would be unfair for the company to benefit by selling them. The IP owner can ask for them to be delivered-up to the court, and can choose to sell or destroy them; destruction is preferred if the IP owner believes the goods are of inferior quality.

In the case of copyright, design right and registered design infringement, there can also be delivery-up of anything designed to make copies of that particular work, such as the mould for an article made of plastic or cast metal.

9.2.2.3 Damages or an account of profits

The infringer can be told by the court to pay a sum of money to the IP owner. The aim is compensation for loss of sales and the amount is often based on the equivalent of a reasonable royalty for the infringing use.

An alternative is for the IP owner to ask for a payment equivalent to the profits made by the infringer, but as accountancy procedures can sometimes show that profit was small or even non-existent, this remedy is rarely requested.

In copyright cases, there can be an award of extra damages for 'flagrant infringement' which means the person responsible for the copying was fully aware of wrongdoing. For breach of confidence, payment of damages is rarely a real compensation, because one cannot make the information secret again.

9.2.2.4 Costs

The successful party usually has an order for costs made in their favour. If costs are awarded but the parties cannot agree among themselves, costs can be set by a Taxing Master of the High Court. The legal representative of the party seeking to reclaim costs provides full details of its activities in preparing for trial including the number of hours spent on meetings, drafting or reading documents, telephone negotiation etc., and the hourly charge for the personnel involved. There will usually also be fees payable to the court and to barristers and other professionals or experts, such as accountants and private investigators. A hearing then takes place when the other side can challenge the figures, and possibly the extent of the work that was done.

The Master sets the cost at the minimum amount regarded as reasonable to have incurred, which is usually between one-half and two-thirds of the actual costs, so this is one area where even the winner loses.

9.2.2.5 Declaration of validity

For registered IP (patents, registered designs or a registered trade mark), the owner can obtain a certificate stating that the right has been held valid. If there is litigation at a later date involving the same registered right, then the owner is entitled to a higher payment of costs.

9.2.3 Where to sue for infringement

The Chancery Division of the High Court has a Patent Court which despite its name is not limited to patents and also hears cases relating to registered designs, registered trade marks, passing-off and copyright. Since 1990 it has also been possible to start an action in the Patents County Court on many of the same types of dispute. In addition, a case can be brought before the Comptroller of Patents in the patent office but the scope of proceedings is narrower than in the two courts and this route is rarely used.

9.2.3.1 Patent Court

In recent years, the High Court judges sitting in the Patent Court (usually only one at a time) have been appointed from the Patent Bar, that is, from barristers with a scientific or engineering training. This is a great benefit when a case involves complex technology.

Cases in the Patent Court must be argued by a barrister, who must be instructed by a solicitor, or, under the direct professional access scheme, directly by a corporate member of an institution.

In patent cases the solicitor is often instructed by a patent attorney, and this also applies for registered design and registered trade mark cases; trade mark agents can instruct in trade mark cases. Having two or three professionals is obviously expensive.

The High Court itself has the right to appoint a professional adviser, and did so in the colour television case mentioned in Section 4.2.4, but such appointments are rare.

9.2.3.2 The Patent County Court

This new court was opened in September 1990 on an experimental basis in Wood Green, London and then moved to central London. It is part of the County Court system but unlike other County Courts, it can deal with cases from any part of the country. It has all the powers of the High Court but a simplified procedure, the aim being to keep smaller-value cases out of the High Court, and to allow small and medium size businesses to bring a case at an affordable cost. It can deal with patent and design cases and with ancillary matters such as copyright infringed when a patent is infringed.

In the Patents County Court, patent attorneys can appear directly or can choose to instruct a barrister. There are therefore fewer professionals involved, and often the patent attorney who drafted the patent specification in the first place can deal with it directly.

9.2.3.3 Court procedure

Both courts are controlled by the same set of Civil Procedure Rules which apply in England and Wales and generally in Northern Ireland; in Scotland the law is slightly different. Proceedings begin with a claim form which sets out briefly the nature of the claim, for example, patent infringement, the remedy sought, for example, an injunction, and a statement of value, how much does the claimant hope to recover. The defendant can file a brief defence. The claimant then has a few weeks to set out the claim in more detail, and the defendant can reply, for example, by alleging that the patent is invalid.

At an early stage, there is a hearing before a judge at which many points are decided, such as which documents are to be disclosed (formerly called 'discovery'), what witness statements will be needed or whether experiments are required. The judge can appoint an independent expert to advise the court but rarely does so.

After further work and further exchanges a trial date is set. At least 4 days in advance the parties must lodge the documents they will use plus a reading guide for the judge which should be short, non-contentious, and preferably agreed by the parties.

A great advantage of this relatively new procedure is that the judge has read the papers in advance and that the oral arguments are limited to really essential points. The time for preparation and for actual trial have been shortened in comparison with the previous system of months or even years before getting to trial.

9.2.3.4 Which court?

A mere decade or so is a short time in the legal system and the pros and cons of the two courts are still argued by the professionals. Some say that the Patents County Court is for simple cases only with no dispute as to facts or need for experts or disclosure of documents. Others appreciate the value of direct involvement of patent attorneys. Whatever the views, both courts provide a far quicker service than 20 years ago.

Both the Patent Court and the Patent County Court can hear cases in camera, that is, with the exclusion of the public, if it is arguable that information to be disclosed should not be publicly available.

9.2.4 Other courts

If one of the parties is not satisfied with the decision of a judge of the Patent Court or Patent County Court, it may be possible to appeal to a higher court, with the permission of the lower court. From both courts, an appeal lies to the Court of Appeal, where three appeal judges will review the decision. If their decision is not accepted by both parties, a final appeal lies to the House of Lords, when five law Lords make their decisions.

In both the Court of Appeal and the House of Lords, majority decisions are possible. If the House of Lords decision is three to two in one direction, while the three Court of Appeal judges and the judge in the lower court came to the opposite decision, it is possible for six judges to have decided against the position which finally applies.

Sometimes, certain aspects of a decision in either court are reviewed by the European Court of Justice (ECJ), based in Luxembourg, although this is not an appeal arrangement and the ECJ cannot override a decision.

9.2.5 Other legal proceedings

In addition to IP infringement cases, there may be 'mini-trials' of IP issues at other times. During the processing of applications for patents, registered designs and registered trade marks, there are often hearings of various degrees of formality at which patent attorneys or trade mark agents put their arguments, usually justifying the grant of a patent or trade mark registration. This is especially common when the European route is chosen for patents, when there may also be an Opposition to the grant of a patent and the case has to be argued by both sides.

In patent cases the inventor may attend and assist but often the professionals deal with the case themselves.

9.2.6 Criminal provisions

Misuse of IPRs can be a criminal offence. Some types of copying of copyright material falls into this category and is covered in Section 2.10. The maximum penalty is 2 years in jail, an unlimited fine or both. Imprisonment has been awarded in a small number cases.

Misuse of a trade mark carries an even higher penalty of up to 10 years in jail. This provision was introduced in the Copyright, Designs and Patents Act 1988 and so far the jail sentences have been in association with other wrongdoings, such as video piracy.

9.2.7 *The professionals*

Different aspects of IP law are dealt with by patent attorneys, trade mark agents, solicitors and barristers. Each has a training relevant to the aspects with which they deal.

9.2.7.1 Patent attorneys

Patent attorneys are engineers or scientists who have trained in the law relating to IP, primarily patents, registered designs and trade marks. Their main work is to deal with all aspects of patents, initially preparing a specification in the area of technology with which they are familiar. They deal directly with the grant of UK and European patents and indirectly through overseas attorneys with patents in other countries. Subsequently, they may be involved in infringement actions.

The term used is 'Registered Patent Agent' (RPA), which means appearing on the UK register of patent agents. There are currently about 1600 such registrations. Chartered Patent Agent (CPA) means a Fellow of the relevant professional institution, the Chartered Institute of Patent Agents; most RPAs are also CPAs. The institute's website is at www.cipa.org.uk. 'Patent Attorney' is also permitted by UK law as an alternative title. 'European Patent Attorney' means qualified to act before the European Patent Office.

Both UK and European attorneys qualify by examination, with several papers covering the relevant area of law and also practical essentials such as the ability to analyse an invention, draft a patent specification and claims, and interpret patent specifications in accordance with legal principles. The European examination also includes technical material in French and German, which must be read and understood to answer the questions.

Patent attorneys can be industry-based, acting only for their employer, or in private practice, acting for any company or individual. Patent attorneys usually also deal with registered designs, and some but not all also deal with trade marks. Some but not all are knowledgeable about other aspects of IP law, such as confidentiality, and some are involved in licensing activities.

9.2.7.2 Trade mark agents

In addition to patent attorneys who act as trade mark agents there is a separate qualification, Registered Trade Mark Agent, for agents who usually have no technical background but often have a qualification in languages or general law. Trade mark agents cannot deal with patents.

Trade mark agents are either industry-based or in private practice. There are about 1000 agents on the register including patent attorneys.

9.2.7.3 Solicitors

Solicitors have a basic training in a very wide range of law, and then usually settle into one field. A small number of private practice partnerships specialise in IP law. Some of them deal with non-technical aspects, such as musical and literary copyright, others with technical areas, such as patent litigation. These days, a few solicitors have a technical training but even without it they are legally permitted to handle all aspects of IP law, including drafting patent specifications.

In patent and other types of High Court litigation, the solicitor is responsible for handling the formal aspects of the case, acting as an intermediary between the patent agent and the barrister who will present the case orally. Some of the specialist firms handling patent litigation have staff with dual patent attorney–solicitor qualifications.

Solicitors in what is known as company-commercial law also deal with confidentiality agreements, contracts such as R&D agreements, licences, company mergers and acquisitions and JV agreements.

Solicitors can be employed in industry or in private practice acting for any client. There are about 50 000 solicitors in England; qualification in Scotland is separate.

9.2.7.4 Barristers

Barristers have the specialised skill of oral presentation of cases in court, and of preparation of court documents. They also give opinions on legal problems, based on their deep knowledge of the law and case law. Equivalents in Scotland are known as advocates. A small number specialise in IP, most are based in London and many have a technical background.

A barrister can be approached by a solicitor or a patent attorney for IP matters, and their clients can also be involved in an interview with a barrister to seek advice or in preparation for litigation; this is known as a conference if with a junior barrister and a consultation if with a QC. A lay client cannot directly approach a barrister unless he or she has membership of an approved institution which has been granted direct professional access.

Patent attorneys, trade mark agents and solicitors are listed in classified telephone directories, and their respective professional institutions provide membership lists, sometimes on a geographical basis; the relevant addresses are given at the end of the book.

9.2.8 Litigation in other countries

Worldwide there are basically two different approaches to the law and litigation. The first applies in the UK and a few other countries, mainly the USA, Canada, Australia and New Zealand. The rest of the world, including all other European countries, take a different approach.

The English approach is based on common law. Legal principles are developed by court decisions over the centuries. In some areas legislation has been formally passed by parliament, but this is far from universal. In litigation, the approach is known as adversarial: the parties each argue their case, often entirely orally, and the judge decides which side has the better argument.

The rest of the world uses the civil law approach. There is much greater emphasis on statute law and less on law developed through court decisions. In litigation, the approach is known as inquisitorial. The aim of the court is to get the full facts, and this is achieved by the judge having close involvement in the preparation of the case. There is far greater reliance on written material, trials are faster and less costly than with the adversarial system. The approach of the Patent County Court, explained above, is much close to the inquisitorial system than any other court in the UK at present.

If a company has patented the same invention in many countries, and a large competitor is infringing in several of them, the owner can decide the best country in which to litigate. The courts in each country may incline to be lenient towards IP, or the reverse; they may have a history of finding patents valid and infringed, or the opposite. These are important factors. Another relevant point is whether the disclosure process applies so that documents and records of the other side can be accessed. In the USA, disclosure is even broader than under English High Court rules. How strongly the court relies on written or oral evidence, and the importance of cross-examination, should also be considered.

In general, for two multinationals in conflict, the outcome in one country will probably be decisive. Polaroid Corporation sued Eastman Kodak Co. for infringement of several instant colour photography patents only in the USA, although in practice (but not in law) the decision applied worldwide. Polaroid won, and in autumn 1990 the court awarded $909 million in damages and lost profits on a 'reasonable royalty' basis.

9.2.9 Management of litigation

It is all too easy for the litigation process to develop its own momentum, and for those handling it to become too involved in the cut and thrust tactics to notice the overall effect. Legal cases can be exciting but they are extremely expensive, and regular overall reviews with the lawyers involved are essential.

It is possible to take out insurance to cover the visible costs of litigation usually as a plaintiff pursuing an infringer, but sometimes also as a defendant, when infringement was inadvertent.

In addition to the visible expenditure on legal fees, there is a huge hidden cost; litigation consumes large amounts of management time and energy in locating in-house information, identifying experts and reviewing mounds of documents. This can easily disrupt the normal running of a business or a department.

A realistic appraisal of what the IP owner or alleged infringer wish to achieve, what the proceedings can actually achieve, and whether it makes commercial sense to continue, must be made initially, and kept under review.

In IP litigation the reason for bringing the case or defending a case aggressively, may vanish as a company develops its business and its interests change. A negotiated settlement is possible at any time, and this should never be forgotten.

9.3 TRIPS – Trade-Related aspects of Intellectual Property Rights

The TRIPS Agreement has a set of provisions relating to general principles for the procedures in member countries for enforcement of IPRs and the remedies available. The objective is to ensure that IP owners can enforce their rights in an effective way.

9.4 Summary of licensing and litigation

Granting permission to other companies to use IPRs can generate useful funds, in the form of a lump sum, a continuing payment or both, and in effect give access to otherwise unattainable markets. Taking a licence to use technology developed by other companies in return for payment can cut out development time and allow a company to move into new technological areas.

Litigation relating to IPRs takes place either in the Patent Court, or in the Patents County Court. The winner of a case can be awarded an injunction to stop the infringing use; payment of damages; delivery-up to the court of the infringing goods and costs.

Chapter 10
Management of IPRs

10.1 Introduction

When the first edition of this book was published in 1994, the term 'IPRs' (intellectual property rights) was becoming generally known, partly by reason of the high profile of IPRs in the Uruguay Round of GATT – the General Agreement on Tariffs and Trade – which was finally agreed in 1995; familiarity was improving but understanding was by no means widespread.

With this edition, the term IPRs is widely seen in the general press, although only the heavyweight end makes any effort to understand the principles, and even journalists in the quality press tend to write about 'patenting your trade mark'.

The terms 'knowledge society' and 'knowledge assets' are fashionable in articles on management, and of course IPRs are a major knowledge asset, with skilled staff being another.

Management journals giving a very broad view of companies point out that 50 years ago a UK company comprised 80 per cent tangible assets, such as plant and machinery, and 20 per cent intangibles; today the position is quite different with intangibles taking a much greater share. The percentage will vary with the type of company; the author has seen figures setting the intangible proportion as 75 or 80 per cent, perhaps for an R&D organisation with relatively little in the way of buildings. WIPO, World Intellectual Property Organisation, sets it at 40 per cent on average on a worldwide basis.

So the basic knowledge of the importance of intellectual property (IP) is accessible, yet in even the largest technical companies, patents are still seen as a necessary evil, with the budget being first in line for reduction; the IP department is seen as a cost centre, not a positive force for revenue generation.

Perhaps we are on the verge of a major change. As pressure mounts to seek all possible sources of income, manufacturing companies have in recent years begun to look at their patents with a view to making money. Patents which are not being used by a manufacturer for protection of their own products can be sold off to generate a lump sum and to remove the cost of renewal fees. Patents which are being used in part, but which can also be used in different applications to those of the owner can be retained but licensed-out to other manufacturers who can launch new royalty – generating products which do not compete with the patentee.

Patent-owning companies which have not yet woken up to the benefits of such an approach may be stimulated by IBM's announcement that it received revenues totalling US$1.7 billion from patent licensing in the year 2000. Texas Instruments realised US$500 million. Total worldwide revenues from patent licensing increased from US$10 billion in 1990 to US$110 billion in 2000, and while most of this will flow to multinationals, some of it will reach small and medium enterprises (SMEs).

So why are patents still regarded in many companies as a drain on resources? Perhaps because no one has yet devised a really simple way to put a value on a patent which keeps competitors out of your market. There are indeed methods of patent valuation, and these are summarised below, but the most realistic is complex to use and still relies on estimates/guesses.

For IPRs to be given their rightful importance the most effective approach for a company of any size is to have a board director responsible for IPRs. The director need not be highly knowledgeable about patent details, but just needs to understand the principles. That way a company-wide view can be taken of benefits against costs and with luck the simplistic view that 'patents cost too much' can be avoided. The large budget needed to obtain and retain patent cover in the overseas territories of importance to the company can be set in the context of the clear control of the market given by a strong patent holding.

The need for an overall IPR 'champion' is not limited to large companies. The lack of knowledge of IP that exists in SMEs, and the cost barriers that SMEs encounter have been referred to earlier. The same principles of IPR management apply whatever the company size.

One possible company policy is to ignore legal protection for its own innovations and hope to be first in the market place each time with its product or service. This is a valid decision, provided it is taken deliberately with a knowledge of the implications – often there will be no right to stop imitators, and no possibility of granting licences. Obviously, the author recommends a more positive IP strategy which leads to more effective and longer term use of innovative effort.

10.2 Overall management of IPRs

In general there are four strands to IPR strategy in a company: generation of new ideas; their protection by use of IPRs; market watch in a broad sense; and trading IPRs by licensing, selling or buying. In all four there are strategic decisions at board level, and daily or weekly decisions at IP management level. Board involvement is most time consuming in the fourth category.

The rest of this chapter will refer to actions by an IP manager as well as the board. This is for large companies; in SMEs the same decisions will be needed but will involve the external patent attorney advising the company, in association with a technical contact at a high authority level in the company so that decisions can be made.

The first three strands will be dealt with in the context of patents, followed by other IPRs. The trading aspects will then be explained for all IPR types.

10.2.1 Generation of new inventions

In a technical business most of the ideas will arise in the R&D department, although any employee in any function may have a useful and creative input to offer. The UK is well known to be good at having ideas and bad at putting them into practice, but encouraging the ideas must be the first step.

10.2.1.1 Board input

Given the close association between R&D and patenting, the first relevant board decision is that of R&D policy. Does the company wish to be a market leader involved in basic research to get ahead and stay ahead of the competition? Or does it want to be a market follower and to engineer developments to give equivalent products or services to that of the market leader? It is a sad but true fact that the second company in a market may show more profit than the first to open up a new area. Nevertheless, companies aiming to be market leaders are likely to invest more heavily in patents than market followers.

The R&D scoreboard produced by the Department of Trade and Industry (DTI), based on the top 700 UK companies and the top 700 companies worldwide, showed that in the year 2004, R&D active UK companies were dominated by pharmaceuticals and biotechnology firms, which have long recognised the value of patent protection for their products; such products often take years to develop and test. R&D expenditure in electronic, electrical and software companies and IT services in the UK was well below international comparators.

The report comments that 'UK companies have a higher cost of funds relative to both sales and R&D than in the USA, Germany, France or Japan. This is driven by higher UK dividends.' All boards must steer their company with this in mind.

Another DTI report, based on research by the Imperial College Management School, states that there is no link between R&D expenditure and company performance. On the other hand, the DTI Innovation Report (see www.dti.gov.uk/innovationreport) sets out in detail statistics about innovation, and points out that SMEs need to learn more about IPRs, with the UK Patent Office starting to give free advice in the second half of 2004.

In the author's view, a vital board decision is to appoint an R&D manager who is capable of encouraging new ideas from staff at all levels – such management skills are outside the scope of this book.

Another area is whether the company can contribute patentable ideas for inclusion in industry standards; with the benefit of a more than average certainty of royalty flow (see Section 4.13), standards-directed work may be particularly encouraged.

Another board decision should specify to whom the patent manager reports. If it is the R&D director, then legal issues will not be understood; if it is the head of legal, then technical issues will not be understood; if it is the finance director, then neither technical nor legal topics will find an educated ear; unfortunately, there are very few patent attorneys who are UK company directors, so there is no perfect reporting line.

Yet another board decision should specify how the IP department is funded. The fashion for close involvement of business units with both R&D and IP funding may give good links between work done and current issues, but that is at the expense of a long-term view. A business unit will think it has better uses for its funds than filing patent applications or for basic research. Central funding of both functions is far more likely to allow long-term planning.

10.2.1.2 Manager decisions on new inventions

Good cooperation between a patent manager and the R&D manager is essential. Working practices which encourage new ideas need to be established, and routes set up so that the ideas reach the patent department for review. How this is achieved can vary. A large multinational may have forms to be filled in and signed, an SME may simply have an open door to the decision-maker, or sometimes vice versa. The system must suit the corporate culture, whatever the company size.

The patent department should be on the circulation list for R&D project reports both interim and final, at an early enough date to allow a patent application to be filed before any sort of disclosure. The patent manager must have the authority to hold up the disclosure if necessary to allow a filing to be prepared.

One way to encourage the flow of ideas is to have a Patent Award Scheme with cash payments for each patent application filed, for each patent granted, and even for each idea submitted. Another way is to have an inventor's tie (most inventors are male because most scientists and engineers are male), or a wall plaque or desk trophy. Internal publishing of lists of inventors' names may encourage others to join in.

10.2.2 Patent protection

10.2.2.1 Board input

Board knowledge of the planned direction of the company and of competitor activity should be the basis for country filing strategy.

Frequently, a company will require patent protection in at least one overseas country, especially for inventions which can be claimed broadly. A claim covering many variations of an invention which is difficult to avoid justifies filing in a larger number of countries than narrow claims which can easily be designed round by a competitor company.

There are four basic factors in deciding where to file. If a company has an overseas manufacturing subsidiary, a patent in that country can protect it. If a company has important export markets, patents in those countries will be beneficial. If the invention is potentially licensable to identifiable countries, patent applications in those countries are essential. Finally, if a competitor has manufacturing facilities abroad, patent applications in the relevant countries may hinder its activities.

The general policy will depend on the type of invention. A product or process which can only be carried out with highly sophisticated plant need only be protected in the relatively small number of countries in which such plant can be established. For a technology which is easier to implement, patent applications in far more countries may be required. Pharmaceutical companies file in many countries but the widest filing sector of all is believed to be the tobacco industry.

Usually, a company will devise several semi-standard lists of countries of interest, depending on the technical area of the invention, but the list needs to be considered in each case.

The overall decision is a commercial one – is the expense justified?

10.2.2.2 Manager decisions

The experience of patent attorneys and the questions they ask the inventor assist a decision on whether each submitted idea justifies a patent application. Is it a major step forward or a small improvement? If small, is it one which solves a production line problem allowing cheaper production so profit can be increased without costs rising? Does it deal with a common customer complaint? Perhaps it is yet another small variation, itself barely supporting a patent, but assisting a spread of cover in a selected area to discourage others from entering it – this is a perfectly valid policy, but an expensive one.

Another patent management decision is when to file. Is a competitor known to be active in this area so an earlier filing might predate their similar filing? Or is the idea just not ready, with more work needed to give substance to a patent specification? If a US application is in view, will the greater detail needed under US law be available in 12 months time?

One option is to file early and then file subsequent applications, possibly at intervals of a few months as the work progresses. There is no publication until 18 months after filing, so these early applications can be abandoned if necessary and the most appropriate specification selected to form the basis of the ultimately granted patent. The cost is not great and updating an earlier filing takes much less of the expensive time of a patent professional than writing a first specification.

A related topic is publication. This is especially important for academics who need publications to establish and maintain their reputation, and usually need to include a deep theoretical analysis. There may be tension between an industrial company providing funding and wanting to delay filing as long as possible, and the academic institution with disclosure at an important conference in view. Whatever the outcome, the patent application must be filed before the publication, whether in writing or by a lecture or demonstration. An application can be prepared in a day or two if necessary, but more time for drafting will give higher quality.

A further balance is between filing a patent application and keeping an invention confidential. If a manufactured product discloses the invention, then a patent will be essential. If the invention is a process which is not detectable in the product, then it can be kept as a trade secret – provided knowledge of it can be highly restricted, and bearing in mind that employees with knowledge of the secret may leave and go to work for a competitor. While they will be under an obligation not to disclose the secret, there is still a risk to the owner.

Once a priority filing is in place, further decisions will be needed towards the end of the convention year – where should applications be filed? Are any non-PCT countries to be included (an example at the date of writing is Taiwan) in which case translations may be needed? If the PCT route is chosen then a major decision is due at 30 months when the international application must transfer into national patent offices, with translations of the full specifications where the working language is not English. Has the invention been implemented successfully or has it failed to reach expectations? Should the country list be maintained or reduced?

At all times of fee payment, a patent manager will be considering whether a patent should be abandoned. National patent offices require renewal fees to keep a patent in force, most are payable annually and they increase with time. A decision not to pay is irreversible so it must not be taken lightly. A short-term cash shortage should not affect the decision and a long-term view needs to be taken. Some patents may be unexploited for years after the investment is made, but if a country is abandoned, or even the whole patent family, then all related investment will be lost.

10.2.3 Market watch

10.2.3.1 Board input

A company board will ensure that it is fully informed of the activities of known competitors and their financial strength. New companies entering the product area can be identified from commercial sources, and the IP department will add the names to the standard name search list.

The board should also ensure there is a system for informing the patent department about new product releases in good time to allow searches to make sure that no third party patent will be infringed. This can be by a reliable flow of reports, but earlier involvement of the IP manager may be better secured by attendance at committee meetings at which progress of new products is reviewed.

10.2.3.2 Manager decisions

A patent watching service covering the patent classification areas relevant to the business is an important source of information on the detail of what others are doing. A few large companies have in-house resources but most use the searching expertise of specialist companies.

Often the results are received each month. The patent department will weed out the irrelevant (it is better to have a few patents listed which are off target than to miss an important one by narrowing the search field; no classification system or search strategy is perfect), and review the remainder.

The questions asked include: is this so close to one of our own applications that we should consider an official objection such as an opposition in the European Patent Office: if this application is granted, will our freedom to improve our product be limited? Would R&D staff be interested in this idea; might it spark a new train of thought and our own inventions? Some companies circulate patent abstracts to R&D

staff as a matter of course, but reading the often strangulated text is never a popular task, whatever the importance.

If a newly published application is of interest or concern, then its progress in the various patent offices is tracked until the form of the granted claim in each is known, and action taken if appropriate.

It is estimated that patent documents contain 70 per cent of the world's accumulated technical knowledge and that most of the information in patents is never published anywhere else or is first published in a patent specification. This vital source of information should never be ignored.

10.3 Management of non-registrable rights

10.3.1 Software

Copyright is by far the most important right.

10.3.1.1 Board input

Having a board member responsible for IP matters applies for all IPRs and not just patents. For a software company the director should ensure that standard software agreements, tailored to meet company requirements, are available, and that there is a policy covering their use. Such agreements will include a non-disclosure agreement (confidentiality agreement) and licences for software use by the appropriate route, for example, a shrink-wrap or click-wrap licence or via a distribution agreement, etc.

Also electronic systems must be in place to record automatically who generates code, and which engineer debugs it, or updates it. This allows proof of copyright ownership. A system must be in place to ensure that all software released to third parties carries a copyright marking – '© IEE 2005' – and any relevant trade mark.

Strong messages must be in place to make sure that all engineers know they must not copy code without permission from the copyright owner. In some companies such an offence results in dismissal.

10.3.1.2 Manager

It is unlikely that the IP manager will even be a separate function except in the very largest software companies, much less an IP attorney, but the manager needs to have basic knowledge of copyright and other IP law plus the management skills to ensure that the policy is followed. This includes use of the standard licence agreements. If changes to these agreements are needed during negotiations, they must be made or approved by a lawyer who not only knows copyright law but also understands the technology to some degree.

If any non-employee writes code, there should be a contract assigning copyright to the company, or a suitable licence agreement for use and exploitation by the company.

All handbooks and instruction manuals should also contain a copyright marking and be under full copyright control of the company.

10.3.2 Other technical rights

For design right and topography rights, as with copyright, the important thing is to ensure that the legal right to use and exploit is owned by the company, either because it originated with an employee or because appropriate contract terms are in place.

For confidential know-how, good systems for dealing with confidential documents should be set up, and staff reminded when necessary about their confidential nature. This is especially important with staff who rarely handle secret material.

10.4 Trade mark policy

The most important decision is to choose a protectable trade mark in the first place, using the principles reviewed in Chapter 6.

Forward planning is needed if a trade mark is to be used abroad, because while a registration is not completely necessary in the UK, filings are essential in some countries if any useful protection at all is to be obtained. Trade marks will usually be the responsibility of a company's marketing department, but some aspects affect use of a trade mark by anyone within the company.

Management systems are needed to ensure that a trade mark is used only in the authorised form. A word mark must always be used with an indication that it is a trade mark, such as use of capital letters, an initial capital letter, putting the word in quotation marks, or following it with 'TM', 'RTM' or ®, although the last two must not be used if the mark is not registered, at least in the home country. A word mark should never be used as a noun, but always as a qualifier for a noun. If a trade mark is in special script or involves a design, then its colour, size and placement on paper also need to be specified and monitored.

There is an obvious need to control the use of a trade mark in brochures etc. but the manner of use in annual reports and even in business letters also needs attention. A company-wide awareness of the proper use of trade marks should also be encouraged. Asking staff to look out for confusingly similar names or logos can give a valuable early warning of misuse by others.

10.5 IPR trading

Intellectual property rights can be bought and sold, licensed-in or licensed-out. The fair price for any transaction is a matter for the parties to negotiate but if a payment is made, taxes are inevitably involved. The type and level of tax will depend on the type of agreement, type of IPR and type of recipient – a company or an individual. Transfers between countries have special tax rules. Expert advice is essential. This section considers only the IPR aspects.

In addition, but usually without payment, a joint venture (JV) or collaborative R&D arrangement can be set up. Licensing is reviewed in Chapter 9.

The split of decision-making between board and IP department will probably be less clear in this area. The board will know the proposed direction of the company, and the IP department will be able to identify the opportunities offered by unused or underused patents, and the threats posed by patent applications filed by competitors, especially those newly entering the same business area.

10.5.1 Product acquisition/asset sale

Sometimes, a product no longer fits with an engineering company's product range but is of interest to another manufacturer. All of the IPRs associated with the product can be bought and sold, as well as stocks, spare parts, work in progress etc.

To transfer the IPRs they first need to be identified, and the risks associated with buying them should be evaluated by the potential purchaser. The process is for the solicitors of the buyer and seller to deal with a list of assets including IPRs, and for disclosure of any known litigation risk.

Listing registered IPRs is fairly straightforward. The records in the company, in the patent attorney's office or at the patent office itself will provide the details. This applies to patents, registered designs and registered trade marks. An unregistered trade mark used as the product name is also easy to identify. The problems more often arise with non-registered IPRs.

For copyright in drawings, the old fashioned drawing office practice by which a draughtsperson initialled and dated a paper drawing, together with similar information on any modifications made to it, gave a good record of authorship. On-screen drawings are usually automatically logged in this way. This record can be matched with employment records to prove the company owns the rights. Usually design right is also included because the design document on which the right is based is very often a drawing. The drawings can also be related to actual products using internal references. The volume of information may be substantial, but the records are usually systematic.

Copyright in software is another matter. A major program will be created by a large team of software engineers and there is no general practice of keeping records. The authors of the program, the engineers who debug the code, and those who update it are all joint authors, each contributing to the creation of a slightly different copyright. All of them should be recorded, for a legally perfect position, and many software generation programs have the facility to do this.

After the IPRs have been identified, a potential purchaser needs to ask whether the rights were generated by employees or by third parties, in which case the contract conditions should be checked. Was there a formal assignment of IPRs or were purchase order forms used with no reference to IP? Does the seller really own the rights?

The sale of assets usually takes place within a tight schedule and detailed investigations may not be possible, such as a thorough consideration of the validity of a patent, but information from the seller's patent attorney about earlier patents cited by various patent offices will allow a preliminary view to be formed.

The next point for investigation is the potential liabilities. Did the seller carry out patent or trade mark infringement searches before launching the product? Does the company have a clear policy on not copying third party designs or software without permission? Is there a risk of litigation, or has litigation already been threatened, or even begun? Will the purchasing company need to be trained in the use of a manufacturing process, or maintenance technique? If so, a technical assistance agreement may be needed, and this will probably limit the time during which the training can take place.

From the answers to all of these questions a potential buyer can evaluate the IPRs being offered, and the risks associated within the purchase of the product rights, and include these factors in the commercial negotiations on price and the final decision.

The whole process is known as 'due diligence'.

10.5.2 Company mergers and acquisitions

Businesses continually expand or contract and companies regularly change ownership. A group may wish to diversify further or expand its market. A subsidiary may no longer fit the developing interests of a group of companies, and may be a better match with a potential purchaser. If a sale is agreed, there needs to be a formal transfer of assets, such as land, buildings, company cars etc., and the same applies to IPRs, which are especially important when technical companies are changing hands. The IPRs may be the most valuable asset, and all the comments above about asset sales apply with a few additional points.

Sale of a company is usually a highly confidential process. Leaked information about it might affect the share price of both the buyer and the seller, and therefore the procedure is protected by confidentiality agreements, and the number of people involved is limited as far as possible.

The information on IPRs provided by the patent attorney will be transferred between solicitors as with an asset sale. The disclosure is usually done within a timescale which is too short to allow a full investigation. A potential purchaser is usually forced to concentrate on the most important product or a vital production process, and accept lower levels of information on IPRs in other areas.

From the formal records and from information provided, a potential purchaser can find out whether the IPRs are owned by the subsidiary which is up for sale or by its parent company. If a parent company owns an important right, checks are made on whether it is assignable with the subsidiary or whether it is used by other subsidiaries and therefore can only be licensed. Assignments from a parent to a subsidiary may be necessary, and confirmation given that unfettered licence rights are available when they are essential to the business being sold.

Lists of IP licences granted to the company by third parties will be needed, together with lists of confidentiality agreements; each must be reviewed to check that it can be assigned to a new owner, or permission to do so requested. Any constraints on the company's freedom to use its own IPRs, such as a licence granted by the company or a JV agreement, need to be assessed.

In general, only an overall view will be obtained and a few potential problems identified. It is sometimes possible to obtain warranties or indemnities from the seller, but in general the purchaser will bear all commercial risks not identified at the time of the acquisition.

Important trade secrets protected by confidentiality also need to be identified, and whether the skill and expertise of any employee is particularly important. If so, that employee may need to be sounded out as to willingness to move to the new company. To protect secrecy this should be done under a personal confidentiality agreement.

10.5.3 Purchase from a receiver

Either a whole company which has gone into receivership or the rights in a product owned by that company can be purchased from a receiver. There are considerable risks associated with such a transaction. A sound IP position is largely dependent on good records and in the case of purchase from a receiver these records may be deficient for two reasons.

The first is that the company may have been poorly administered, so the records never existed in the first place or were incomplete. The second is that immediately after appointment the receiver removed important documents to stop them from being destroyed. Most receivers are accountants and therefore tend to consider only financial records as 'important documents'. Records of IPRs or the company's technology will always be regarded as less important.

If such records are left on company premises where the company is still continuing to function, it is not unknown for some of them to vanish before a potential purchaser is identified. The records which could establish ownership of the technology of the company may not be available. A potential buyer, knowing the company before its financial difficulties reached a critical stage and knowing its product or services, may be disappointed with the IPRs available for sale.

10.5.4 IPR trading – general

All transfers of ownership, whether a company or product acquisition, or from a solvent company or one in receivership, need formal transfers of IPRs. These should be recorded at the appropriate registry at the patent office. If not, and if there is a subsequent transfer, such as a second sale by an unscrupulous owner, the second registered assignment takes precedence over the earlier unregistered change of ownership.

While the formalities of IPR transfers will be handled by either a patent attorney or a solicitor as appropriate, the client should always check that the technical aspects have been fully covered.

10.6 Collaborations

10.6.1 Collaborative R&D

If two or more parties wish to benefit from the grants for R&D available from the Department of Industry (such as the LINK scheme) or from the EU (e.g. EUREKA), then they must comply with the conditions of these grants. Some of the conditions concern IPRs.

The stated aim is usually that the technology derived from the funding must be used for the benefit of UK or EU industry. To achieve this, the conditions are that the results of the research should be exploited, and the first step is to protect the innovations by IPRs.

A fundamental condition is usually that all of the information is supplied to all of the collaborators. The parties are often required to identify background IPRs, that is, pre-existing rights and information, and to make them available to the other parties to carry out the collaborative work. They are often also used for exploitation of the results – sometimes in return for a royalty payment.

When there is an academic partner, it may be a condition that the industrial partners own the IPRs and pay a royalty to the educational establishment when the results are exploited, whether by licensing or by internal manufacture.

The guidelines of the various schemes vary as to the level of information given on IPRs. They can be very sketchy or quite detailed but at the very least there will be a requirement that arrangements are made for IPRs.

For each scheme a potential collaborator will need to check the restrictions placed on IPRs by that particular funding scheme and decide if they are acceptable. In addition, arrangements will need to be negotiated with the other collaborators. If your company is going to own the IPRs arising from the collaborative work, a check should be made on any restrictions on the use you can make of them – another collaborator may object to your using them in a particular technical field – and negotiate on this basis. If your company is not going to own the IPRs, you will need to negotiate appropriate freedom to use them. Industrial collaborators in particular need to check what background information needs to be made available. A subsidiary in a group of companies should ensure that the rights can be licensed and check whether internal permission is required.

In general, all collaborators will need to think well ahead about the possible routes for exploiting the results of the collaborative R&D, and negotiate appropriate arrangements with the other collaborators. There may be an obligation, on paper at least, to disclose the results not only to other collaborators in this project, but to others receiving funding under the same scheme. It may be possible to protect the position by excluding disclosure to particular named companies such as a strong competitor.

Within the framework of each collaborative R&D funding scheme, it is usually possible to negotiate a position which is acceptable to everyone, but engineering managers need to give careful thought to the restraints and future freedom before committing the company, and to bear EU regulations in mind (see Section 8.4).

10.6.2 Joint ventures

When two or more companies consider setting up a JV, checks on IPRs similar to those made before beginning collaborative R&D are valuable to establish the freedom of each joint venturer to contribute IPRs, and the freedom each will need to use IPRs arising. The main constraint will be EU regulations, particularly those relating to R&D agreements (see Section 8.3.5).

10.7 Valuation

10.7.1 Valuing patents

With the rise in general interest in intangible assets, there has been a parallel rise in interest in valuing, or trying to value, those assets. The several possible methods are being themselves evaluated academically. Software is being written to automate those evaluations, but at the date of writing there is no front runner, or even a small number of programs of greater popularity than the rest.

Some companies have shown an interest in putting IPRs on their balance sheets, but in the UK the Accounting Standards Board has expressed doubt that intangible assets can be properly valued.

As to the methodology, there are three main approaches – cost based, market based and economic based.

The cost-based approach considers how much it has cost to get the patents granted, plus how much has been paid in renewal fees. The disadvantage is that some patents meet few objections in patent offices, so costs are low, but the commercial value can be high. Conversely some applications, even when well drafted and with little prior art, meet major objections and are therefore costly but still of low commercial value.

The market-based approach depends on the use of publicly available values for similar IP, for example, when it changed hands or was litigated and damages were awarded. The main problem here is that litigation is rare, the price of a sale of IP is usually kept secret, and in any case the chance of the available information being any sort of match to that in question is very small indeed.

In the economic-based method, cash flow associated with the right, such as royalty income, is identified and used to estimate future revenue. However, it is not always easy to link a royalty to a specific right, more usually it is linked to a bundle of rights of different maturities, so prediction is not easy.

Perhaps the best that can be said is that with more widespread use of valuation, a greater consistency will eventually develop. For an individual company, at least the annual application of the same technique allows trends to be seen.

A similar argument applies to the valuation software. Input values include the answer to questions such as 'How broad are the claims?', 'Is the patent granted or pending?', 'Is the technology unique or does it depend on a licence-in?'. Guesses are entered of marketing value, of the cost of future development etc.

Certainly all the factors listed in the many versions cover questions which need to be asked, and certainly use of the same software each year allows the development of a patent portfolio to be evaluated on a consistent basis; but for valuation of a single patent the various guesses add up to a great deal of uncertainty.

10.7.2 Valuing trade marks

Valuation of trade marks predated that of patents and is more convincing. By repeated use of the same methods, brand valuation companies publish annual estimates of the most valuable brands in the world. Coca Cola® comes top, with Microsoft® and IBM® next.

An early use of trade mark valuation was when the Swiss multinational Nestle made a hostile takeover bid for the Yorkshire-based confectionary company Rowntree. As part of the arguments for raising company value and therefore the price to be paid, the value of all the Rowntree trade marks such as Aero, Smarties and Yorkie was put forward. In the end, Nestle paid £0.5 million for tangibles and £5.5 million for intangibles, mainly the trade marks.

10.8 Encouraging innovation

It has been stated that innovation is generally agreed to be the life blood of successful industry. As measured by patent applications, and particularly by the technological balance of payments, the UK is tending to lose ground. A recent House of Lords enquiry has pinpointed structural and performance features which characterise innovative companies. The enquiry resulted in three important recommendations:

'To be open to innovative developments and technologies from any source'. It is hoped that this book has demonstrated that innovations can be made by engineers at any level and in any function in an engineering company, and that in many cases an appropriate legal right is available to protect that innovation.

'To learn about the activities of competitors'. Such knowledge can allow a company to avoid being sued, and the publicly available information can allow a company to direct its own innovative efforts appropriately. Technology developed by others is often available for use under a licence agreement.

'To cultivate a positive and enthusiastic attitude to innovation, led by the chief executive and board'. While a top-down approach is the most effective one, awareness of the benefits of innovation and knowledge of the mechanisms for protecting it should be company wide. The aim of this book is to assist the process.

USEFUL ADDRESSES

Chartered Institute of Patent Agents
95 Chancery Lane
London WC2A 1DT

Tel: 020 7405 8450
Fax: 020 7430 0471
Email: mail@cipa.org.uk
Web: www.cipa.org.uk

Institute of Trade Mark Attorneys
Canterbury House
2-6 Sydenham Road
Croydon
Surrey CR0 9XE

Tel: 020 8686 2052
Fax: 020 8680 5723

The Patent Office
Concept House
Cardiff Road
Newport
S Wales NP10 8QQ

Tel: 01633 814 000
Fax: 01633 814 444
Web: www.patent.gov.uk

The British Library
96 Euston Road
London NW1 2DB

Tel: 0870 444 1500
Web: www.bl.uk

The Library has copies of all UK and foreign patent specifications, design registrations, and extensive IP reference material.

Public Libraries in Aberdeen, Belfast, Birmingham, Bristol, Glasgow, Leeds, Liverpool, Manchester, Newcastle-upon-Tyne, Plymouth, Portsmouth, Plymouth, Sheffield and Swansea also have copies of UK patent specifications

The Licensing Executives Society in Europe
LES Britain and Ireland
Gill Moore
Northern Networking
1 Tennant Avenue
College Milton South
East Kilbride
Glasgow G74 5NA

Tel: 01355 244 966
Email: gill@glasconf.demon.co.uk

British Technology Group
10 Fleet Place
Limeburner Lane
London EC4M 7SB

Tel: 020 7575 0000
Fax: 020 7575 0010
Email: info@btgplc.com
Web: www.btgplc.com

Design Council
34 Bow Street
London WC2E 7DL

Tel: 020 7420 5200
Fax: 020 7420 5300
Web: www.design-council.org.uk (which has a list of organisations offering advice on design and related matters)

Advice may also be obtainable from:
Chambers of Commerce
Regional offices of the Department of Trade and Industry
Local Innovation Centres

Index